A Practical Guide to Cooperative Learning in Collegiate Mathematics

MAA Notes Series

The MAA Notes Series, started in 1982, addresses a broad range of topics and themes of interest to all who are involved with undergraduate mathematics. The volumes in this series are readable, informative, and useful, and help the mathematical community keep up with developments of importance to mathematics.

MAA Notes

11. Keys to Improved Instruction by Teaching Assistants and Part-Time Instructors, *Committee on Teaching Assistants and Part-Time Instructors, Bettye Anne Case,* Editor.
13. Reshaping College Mathematics, *Committee on the Undergraduate Program in Mathematics, Lynn A. Steen,* Editor.
14. Mathematical Writing, by *Donald E. Knuth, Tracy Larrabee, and Paul M. Roberts.*
16. Using Writing to Teach Mathematics, *Andrew Sterrett,* Editor.
17. Priming the Calculus Pump: Innovations and Resources, *Committee on Calculus Reform and the First Two Years,* a subcommittee of the Committee on the Undergraduate Program in Mathematics, *Thomas W. Tucker,* Editor.
18. Models for Undergraduate Research in Mathematics, *Lester Senechal,* Editor.
19. Visualization in Teaching and Learning Mathematics, *Committee on Computers in Mathematics Education, Steve Cunningham and Walter S. Zimmermann,* Editors.
20. The Laboratory Approach to Teaching Calculus, *L. Carl Leinbach et al.,* Editors.
21. Perspectives on Contemporary Statistics, *David C. Hoaglin and David S. Moore,* Editors.
22. Heeding the Call for Change: Suggestions for Curricular Action, *Lynn A. Steen,* Editor.
24. Symbolic Computation in Undergraduate Mathematics Education, *Zaven A. Karian,* Editor.
25. The Concept of Function: Aspects of Epistemology and Pedagogy, *Guershon Harel and Ed Dubinsky,* Editors.
26. Statistics for the Twenty-First Century, *Florence and Sheldon Gordon,* Editors.
27. Resources for Calculus Collection, Volume 1: Learning by Discovery: A Lab Manual for Calculus, *Anita E. Solow,* Editor.
28. Resources for Calculus Collection, Volume 2: Calculus Problems for a New Century, *Robert Fraga,* Editor.
29. Resources for Calculus Collection, Volume 3: Applications of Calculus, *Philip Straffin,* Editor.
30. Resources for Calculus Collection, Volume 4: Problems for Student Investigation, *Michael B. Jackson and John R. Ramsay,* Editors.
31. Resources for Calculus Collection, Volume 5: Readings for Calculus, *Underwood Dudley,* Editor.
32. Essays in Humanistic Mathematics, *Alvin White,* Editor.

These volumes can be ordered from:
MAA Service Center
P.O. Box 91112
Washington, DC 20090-1112
800-331-1MAA   FAX: 301-206-9789

©1995 by the Mathematical Association of America

ISBN 0-88385-095-8

Library of Congress Catalog Number 95-76289

Printed in the United States of America

Current Printing

10 9 8 7 6 5 4 3 2

A Practical Guide to Cooperative Learning in Collegiate Mathematics

Nancy L. Hagelgans
Ursinus College

Barbara E. Reynolds, SDS
Cardinal Stritch College

Keith Schwingendorf
Purdue University North Central

Draga Vidakovic
North Carolina State University

Ed Dubinsky
Purdue University

Mazen Shahin
College Misericordia

G. Joseph Wimbish, Jr.
Huntingdon College

MAA NOTES NUMBER 37

Published by
THE MATHEMATICAL ASSOCIATION OF AMERICA

Preface

This guide can greatly help readers introduce cooperative learning in their undergraduate mathematics classes. Instructors who have tried some group activities as well as those who have not been involved at all with cooperative learning will find detailed, useful discussions on every aspect of cooperative learning. The book reflects the extensive experience of the authors as well as that of over forty colleagues who responded to a Survey on Cooperative Learning. Throughout the book cooperative learning is related to educational research results, which are clearly explained in Chapter 2.

Cooperative learning, as used by the authors, involves students working in heterogeneous groups, usually assigned for the duration of the course. Students become responsible for each other's learning since the cooperative spirit permeates every facet of the course: homework, computer laboratory assignments, classes, and even some tests and quizzes.

The book includes directions for organizing students into groups as well as complete descriptions of what these groups might do once they are formed. Examples of group problems and group test questions for various mathematics courses illustrate the work that can be expected of students in cooperative learning groups. Methods for monitoring groups and dealing with problems that may arise are presented. The whole question of assessment of students' work is addressed.

In addition to describing their own methods, the authors include a chapter summarizing some forms of cooperative learning that others have used. There is also an extensive bibliography with many annotations.

We hope this book will provide a valuable resource for any instructor who is considering cooperative learning groups in an undergraduate mathematics class.

A project of this nature does not grow in a vacuum. We have been supported and encouraged by our colleagues at our respective institutions throughout the nearly three years it has taken to bring this project to life. The respondents to the Survey on Cooperative Learning distributed in the fall of 1992 shared experiences that helped to shape this work.

We are grateful for the ongoing feedback that we have received from our students over the years as we have gradually developed the model of cooperative learning that we describe here. And we hope that the methods described here will be used to the benefit of future generations of students.

We would especially like to thank Warren Page for his encouragement and helpful comments after reading an early draft of this manuscript.

Our work on this project has been partially supported by grants from the National Science Foundation; grants #USE-9053432 and # DUE-9450750. The recommendations and conclusions expressed here are solely those of the authors and not of the National Science Foundation.

Finally, we would like to thank those closest to us, family and friends, who have supported and encouraged each of us in many ways.

Nancy L. Hagelgans
Barbara E. Reynolds, SDS
Keith E. Schwingendorf
Draga Vidakovic
Ed Dubinsky
Mazen Shahin
G. Joseph Wimbish, Jr.
March, 1995

Contents

Chapter 1

What Is This Book About?

This book is about *cooperative* learning in *collegiate* mathematics classes. It reflects the collective experiences of the seven authors and forty-two respondents to a Survey on Cooperative Learning that was conducted in fall, 1992. We offer this as a practical handbook for others who might want to try to use cooperative learning as a way of organizing classes of adult learners — both traditional-aged undergraduates and older, returning adult students.

1.1   How did we get involved in this project?

As calculators and computers have become more available, mathematics educators at all levels have been experimenting with ways to use them to enhance student understanding of mathematical concepts. As students have worked with these tools, educators have observed increased interaction among the students. Some instructors began to notice that students were remembering more and apparently understanding important concepts more deeply. These educators began — informally at first, then in more structured ways — to use cooperative learning as a component in their courses.

Research supports these hunches. At conferences, people began to talk informally about their experiences using cooperative learning groups in their classrooms. Articles on cooperative learning started to appear in such journals as *Educational Leadership, Journal for Research in Mathematics Education, Mathematics Teacher*, and *Black Issues in Higher Education*. Papers on using cooperative learning groups in the classroom were being presented at various regional and national conferences.

In the spring of 1992, seven of us — Nancy Hagelgans, Barbara Reynolds, Keith Schwingendorf, Draga Vidakovic, Ed Dubinsky, Mazen Shahin, and Joe

1

Wimbish — began sharing with each other our excitement about what we saw happening among the students in our classrooms. We had each been using cooperative groups for several years in such classes as precalculus, calculus, discrete mathematics, and abstract algebra. While we had started by using computer activities, we suspected that the conversations and interactions among our students were probably at least as effective as the computer activities in helping our students to understand and learn mathematics. At that time, most of the literature on cooperative learning seemed to be based on experiences of teachers in elementary and secondary classrooms. Even now, relatively little has been written on cooperative learning in collegiate mathematics.

The seven of us have tried a variety of approaches to implementing cooperative learning in our classrooms. For us, some things have worked better than others. We are often asked about what we are doing. We think we have something to offer from our experience.

1.2   Survey on cooperative learning

As the seven of us shared our experiences and reflections, we realized that we could learn a lot from each other. We were asking each other the same questions we had been asked by other colleagues: What are you doing? Why are you doing it? How do you organize your classes so that small groups of students cooperate effectively? What do your students do in class? How do you assess student learning? ... And the list goes on.

Each of us also knew others who were using some form of cooperative learning groups in collegiate mathematics classes. In the summer of 1992, we developed a survey that we distributed to about eighty of our colleagues the following fall. We received forty-two responses. These responses to our survey were often accompanied by the students' written remarks taken from course evaluations. In reviewing the survey responses, we realized that there are many variations on how cooperative learning groups are structured, how they function, and how they are used in collegiate mathematics classes.

Throughout this book, the collective "we" is used to refer to the personal experiences of the seven of us, and to the extensive conversations that we have had among ourselves. When referring to ideas and experiences common among the respondents of our survey, we use "the respondents."

1.3   What is "cooperative learning"?

Cooperative learning seems to have a wide range of "definitions." Even among the seven of us, there was diverse opinion about what we mean by cooperative learning. In one long (and rather heated) conversation, the group of seven nearly became a group of five or six because some of us did not see ourselves in the definition offered by the others. We kept talking until we could find common ground, those elements that the seven of us could agree that we were using in our classrooms. During the process of negotiating the meaning of cooperative learning, we considered our own experiences, the experiences of others, and our understanding of Piaget's theory of how people learn. More about Piaget's theory will be given in the following chapter (see Section 2.3).

In talking to others about their experiences with cooperative learning, the authors have found that implementations of cooperative learning run the gamut from very loosely organized to highly structured classroom settings. We think that it is important to clearly state what we mean (and what we don't mean) by cooperative learning in this book. We have identified several characteristics that we think must be evident for the groups to be called "cooperative learning groups."

1.3.1 Significant amount of group work

When we describe a course or classroom setting that is using cooperative learning groups, we mean that a significant amount of the required work for the course — in the classroom, in the computer lab, at home — involves the students in regular interaction among their group members. That is, the course is structured so that the students are involved in or participate in *a significant amount of group work*. We expect that the course is structured so that students need to communicate with each other regularly and often. This communication within the groups includes reflection on mathematical ideas and discussion of alternative approaches to problem solving.

The course activities are designed so that the students are encouraged — and required — to become involved in the kind of discussion that leads to multiple approaches to solving the problems that are posed. In trying to communicate ideas to others in their group, students must clarify their own thinking about a problem or concept. This discussion should be occurring regularly enough and at such a level that students begin to recognize and correct their own errors — both minor errors in computation and more fundamental errors in mathematical reasoning.

Some instructors have introduced group work into their courses by having the students work together on one group project while continuing to teach in an otherwise traditional manner. This is not what we mean when we talk about cooperative learning.

1.3.2   *Esprit de corps* among members of a group

We expect an *esprit de corps* to develop among the members of each group; that is, we expect the students to develop a feeling of belonging to a group — belonging to *this* group rather than to *that* group. Typically we ask the group members to choose a name for their group. Activities that require the group to turn in work are identified by the group name rather than the names of the individuals. When individuals have questions, they may be referred back to their group or asked what the group has tried so far.

1.3.3   Mutual responsibility among group members

Course activities are designed so that a cooperative spirit permeates every facet of the course. Group members are expected to be in some way *responsible for each other*. If a student is absent, he or she is expected to get notes from someone in the group. Some tasks are assigned — and some of the evaluation is done — in such a way that all members benefit when all members perform well. Sometimes a single group grade is given for a project; sometimes individuals get two grades, their own individual grade and the average of the grades of their team mates. Another strategy is to have the students take a test individually, and then meet in their groups to do the same test again using their collective knowledge; in most cases, the group score is higher than any of the individual scores. Some activities are designed so that the *collective knowledge of the group* (which is more than the "sum" of the knowledge of the individuals in the group) is tapped.

While an individual's performance can help the group, an individual's non-performance does hurt the group — at least insofar as the whole group is diminished if one member disengages from group activities. After all, isn't this the way things work in "real life" — in families, in work-related project teams, in sports? If one member of the family or team excels, all benefit; if one member "messes up," all are penalized by the natural consequences of one person's mistake.

1.3.4   Stable groups

In order to foster the development of this *esprit de corps*, we think that it is essential that *assignment to groups is permanent*, preferably for a whole semester or a significant part of the semester. Learning to work together — learning how to use group interaction to learn mathematics — is a process. Students need to know that we expect them to learn how to use the group. Even students who are already comfortable working in groups need to learn how to work within the dynamics of the particular group to which they now belong.

1.3.5   Evaluation of group work

Because we seriously believe that students really do learn better by working within the group than by working alone, we think that we must *include group work in the evaluation process.* At least a part of the grade for the course is based on work that students do together. Among ourselves and among the respondents to our survey, this is done in a variety of ways. Some of us give group grades on homework; some have used some form of group work on the tests. (We go into more detail about this in Chapter 5.) In the long run, we know — and our students know — that the "test" reflects everything that we think is essential. So to preach the value of cooperative group work but to do all the evaluation in a competitive, individual setting would be to undermine what we say about the importance of the group.

1.4   Essential components of cooperative learning

In summary, when we speak of *cooperative learning* all of the following components must be present:

- a significant amount of the work of the course is done in cooperative groups,

- a positive *esprit de corps* exists within the groups,

- team members share a feeling of mutual responsibility for each other,

- group membership is permanent and stable, and

- group work is included in the evaluation process.

Throughout this book, we will be sharing many of our own experiences as well as experiences shared with us by the respondents to our survey. While there are other kinds of structures that provide opportunities for students to interact and cooperate with each other, when we talk about cooperative learning in this monograph, we expect all of these principles to be operative. The students participate in permanent groups in which members share responsibility in the learning process. The classroom environment is structured so that students interact regularly within their groups. The evaluation process contains some group work (including group tests). We expect the classroom environment to be conducive to the development of an *espirit de corps* among the members of each group, and to be structured so that the actual process of learning involves communication within small groups.

1.5　A brief overview of this book

Chapter 2 gives a theoretical overview of learning theory research in cooperative learning. Some results from a study of a group of undergraduate mathematics students are presented, and the benefits of cooperative learning are discussed.

Chapters 3, 4, and 5 address practical issues.

In Chapter 3, we present some of our strategies for forming cooperative groups. We reflect on our experiences, and share our observations of what has worked well for us and what has worked poorly.

"What do the groups *do*?" is a frequently asked question. In Chapter 4 we discuss various activities, tasks, and exercises that we have used in our classes. These include activities done in the problem sessions and computer laboratories before a topic or concept is formally introduced, tasks done in the classroom, homework exercises, and group preparation for examinations.

If we really believe in cooperative learning (and we do!), the methods we use to assess students' progress in the course must reflect our commitment to working cooperatively. The challenge, then, is to find an appropriate balance between affirmation of the group process and individual accountability. Chapter 5 offers some practical ideas on how to assess the learning that (we hope) is really going on in the minds of our students.

Groups, like families, can be more or less "functional." In Chapter 6, we discuss ways that groups can be more or less productive. When a group has more non-productive time than productive time, we call the group "dysfunctional." In this chapter, we include strategies we have used when groups in our own classes have been dysfunctional.

In Chapter 7 we reflect on what our students and colleagues have told us about their experiences with cooperative learning. We offer some observations on the comments that were written by the respondents to our survey. (A compilation of all the responses to this survey is included in Appendix B.)

In Chapter 8, we briefly survey how others are using cooperative learning.

The bibliography offers a broad sampling of cooperative learning literature. Many of the items in this bibliography are annotated with abstracts selected from the ERIC database.

Finally, the appendices include a variety of material that may be helpful to you as you consider using cooperative learning in your classroom. Because much of our discussion refers to it, we include a copy of the original questionnaire that was sent to our colleagues in fall, 1992 (Appendix A). While we summarize the comments made by instructors in the survey in various places throughout this book, we include a compilation of all the responses we received on the survey, almost verbatim, as they appeared on the forms returned to us in Appendix B. We also include samples of forms that some of us have used in our classes: a student information questionnaire used for forming groups (Appendix C), and a class participation form used for daily feedback to the instructor (Appendix D). We've included samples of the statements in course syllabi that tell how cooperative learning is used in the class (Appendix E), as well as some sample grading schemes (Appendix F).

1.6 How to read this book

It is not necessary to read this entire book in a "linear" way, from cover to cover. A reader with a bit of experience in cooperative learning might browse through various parts of the book using this brief overview and the table of contents as a guide to topics of particular interest. For the reader who is new to cooperative learning, most of the nitty-gritties of classroom management are contained in Chapters 3 - 6. If you haven't yet decided whether you really want to try cooperative learning in your own classroom, you might begin by reading Chapter 7 in which we reflect on our experiences and the experiences of our students in a qualitative way. If you've had some experience with cooperative learning, the theoretical background given in Chapter 2 helps to explain why or how cooperative learning actually facilitates the learning of mathematics.

1.7 The experience of working cooperatively

What is it like for seven very busy people living in six different states to work cooperatively on a project like this? (When we started this project, we were living in Alabama, Indiana, Massachusetts, North Carolina, Pennsylvania, and Wisconsin.) Perhaps our experience is not all that different from the experiences of others who collaborate in research as well as in teaching.

We had an initial meeting in June, 1992, which only four or five of us were able to attend. We wrote the survey questionnaire making use of the Internet and fax machines since not all of us were on the Internet. Even now, only six of us are on the Internet. We agreed to a tentative outline of the book via the Internet; the outline has been modified at least twice. Our first complete draft of the book came together over Thanksgiving weekend, 1993. We had our first face-to-face meeting of the entire group at the Joint Mathematics Meetings in Cincinnati in January, 1994. At one or the other time, each of us has fallen a bit behind in our work on this project. As can be expected, about the time one person fell behind another became enthusiastic and wanted to see "real progress" being made on the project. At times our sense of connection has been a real source of excitement, and we have marvelled at what we have seen grow by our collective efforts; at other times, we've each felt frustrated that someone else hasn't seemed to be contributing in the same measure as the rest of us.

Each of us, as a member of this cooperative team, has had first-hand experience of the frustrations our students sometimes mention. We have also seen this project take on a life of its own. While there are things in this book that each one of us has assumed direct responsibility for, the experiences reported here are our collective experiences. No one of us could have written this book alone. Had one of us tried to do so, it would have been a different book.

The breadth of experience reflected in these writings is bigger than any one of us. Each of us has been stretched to see things from the point of view of colleagues whose institutions are very different from our own. We believe that the collective experience that we are able to share with you here is greater than any one of us could have offered individually.

Does cooperation work? We believe that it does. We leave you, the reader, to make the final evaluation for yourself as you read this book and observe your students interacting in their groups.

1.8   Chapter Summary

This book is the result of the collaborative efforts of the seven authors, who have been using cooperative learning in collegiate mathematics courses for a number of years. In 1992, we wrote to other colleagues who also were using cooperative learning in collegiate mathematics courses, and we received forty-two responses to our Survey on Cooperative Learning.

Although the term "cooperative learning" has been used to describe a variety of learning environments, in this book we have taken a focused perspective. Cooperative learning as we use it in this book includes all of the following components:

- Students participate in permanent, stable groups.

- A significant portion of the required work of the course is done in groups — in the classroom, in the computer lab, and in homework.

- The evaluation process includes group work.

- A positive *espirit de corps* is fostered among the members of each group.

- The classroom climate is such that a spirit of mutual responsibility develops among group members in the learning process.

In later chapters, we offer some theory and many practical suggestions. We answer some of the questions that we are most frequently asked — Why are you using cooperative learning groups in your classes? How do you form the groups? What do the groups do? How do you assess your students' progress? What do you have students do on a group test? How do the groups function? What kinds of things can go wrong within the groups? How do you handle problems within groups? We include a chapter that surveys how others are using cooperative learning, and an extensive annotated bibliography. The appendices contain our survey questionnaire as well as some practical forms and information that you might use as you consider implementing cooperative learning in your classroom.

Chapter 2

Why Use Cooperative Groups?

A growing number of educators who have used cooperative learning in collegiate mathematics classrooms have reported positive experiences. They say that their students seem to have a deeper understanding of mathematical concepts, and remember what they study for a longer time (e.g., from one semester or course to the next). They say that their students show greater interest in attempting to solve problems — even challenging problems. These instructors also report that they have renewed personal enthusiasm for teaching.

2.1 Not all small group learning is "cooperative"

Numerous mathematics educators and researchers in mathematics education have been critical of the predominant use of the lecture as a primary instructional method. Many commentators advocate finding some viable alternative to the lecture method. They argue that whatever learning occurs just from lectures is at best passive and tends toward memorization. While memorization as such is not a bad thing, the understanding it fosters is what Skemp (1987) calls instrumental as opposed to relational understanding. Others (Schoenfeld, 1990; Krantz, 1993) argue that little or no learning can be directly attributable just to listening to a lecture.

Of the many alternatives suggested, one of the most frequently mentioned is some form of small group study. Often the small group form of instruction is labeled cooperative learning. Unfortunately, working in small groups does not

11

guarantee that any learning or cooperation takes place. In fact, it is not always clear even from the context what is meant by the term *cooperative learning*. Although some researchers and educators report that little or no learning has taken place, and others report no difference from other methods, it is often not clear what form of "cooperative learning" has been used. In particular, it is not clear whether the methods used in those studies are comparable to what the authors of this monograph intend.

The first chapter provides a set of guidelines explaining the meaning of cooperative learning used in the present context. In this chapter we explore some of the reasons for using cooperative learning, mention some of its benefits, and cite some of the evidence supporting the use of cooperative learning groups. We describe some of the pitfalls occasionally encountered using cooperative learning. We discuss some issues of learning theory that may help both instructors and students to optimize the benefits of cooperative learning.

2.2   Research on the learning of mathematics

2.2.1   Attitudes and beliefs

When students arrive at college, they bring with them a variety of attitudes and beliefs. They have definite opinions concerning the nature of mathematics and the reasons for studying it. They think they know how mathematics should be taught as well as what roles students and instructors should play in a mathematics classroom.

Researchers (e.g., Dubinsky, 1989a, 1989b; Garofalo, 1987; Garofalo and Lester, 1985; Schoenfeld, 1985; Schwingendorf, Wimbish, and Hawks-Hoover, 1992; Wimbish, 1992) list some of these attitudes and beliefs. For example, Dubinsky identifies four beliefs or attitudes concerning the learning of mathematics: one can learn mathematics spontaneously, inductively, constructively, or pragmatically. These beliefs about learning mathematics correspond to four categories of belief about the nature of mathematics. Mathematics is:

- a body of knowledge already discovered that must be passed on to future generations by transfer from the teacher's mind to the students' minds;

- a set of techniques for solving standard problems that must be practiced until mastered;

- a collection of thoughts and ideas that individuals and groups of individuals have created and constructed and that students might be expected to also construct; and

- a set of applications that stress only the power of mathematics to describe, explain, and predict.

Other researchers comment on the importance of students' beliefs and attitudes (e.g., Garofalo, 1987, 1989; Garofalo and Lester, 1985; Schoenfeld, 1985). For example, Garofalo identifies five commonly held beliefs that students have concerning mathematics. These beliefs include:

- Almost all mathematics problems can be solved by the direct application of the facts, rules, formulas, and procedures shown by the teacher or given in the textbook.

- Mathematics textbook exercises can be and must be solved by the methods presented in the textbook.

- Only the mathematics to be tested is important and worth knowing.

- Mathematics is created only by the mathematical genius; others just try to learn what is handed down.

- Mathematics problems have only one correct answer, and these answers are obtained by using a step-by-step algorithm.

Schoenfeld (1985), however, observes the following contradiction. Students simultaneously claim that "mathematics is mostly memorization," and "mathematics is a creative and useful discipline in which they learn to think."

A central issue in mathematics teaching and learning is one's beliefs and attitudes toward mathematics. Many instructors who use cooperative learning hold the belief that mathematics is learned constructively, that mathematics consists of thoughts and ideas that individuals and groups of individuals have created and constructed, and, most important, students should be helped and expected to do the same.

2.2.2 Effect of cooperative learning on attitudes

Research and anecdotal evidence suggest that students in classes using cooperative learning, as defined here, can develop a more positive attitude toward themselves and mathematics. Working exclusively with non-major mathematics students, Wimbish (1992) observed changes in attitudes and beliefs. For example, students, who at the beginning of the course were very tentative about their abilities, thought they were doing as well as others in their group by the end of the course. Individual students expressed a willingness to help themselves or to help others. After participating in this form of cooperative learning, students were more willing to test their ideas and more willing to explore new and better ways of solving old problems.

Wimbish observed also that some old attitudes persisted after one semester of work in cooperative groups. For many students, "formula memorization" remained the essence of mathematics, and there was still a strong tendency for the students to link understanding with memorization. The students, on the other hand, did express that they tried to explain a concept to a member of their group, tested their understanding of a concept by trying to explain it to another student, and practiced to improve their computer skills in using the computer outside scheduled laboratory time.

In preliminary results from a longitudinal study over a two-semester course in calculus for students of management, social and life sciences, Schwingendorf and Wimbish (1994) report other changes in the attitudes and beliefs of students. These students displayed a more positive attitude toward mathematics and to their ability to solve problems. They showed a willingness to talk about doing mathematics as a collaborative approach to problem-solving. These students were less dependent on an instructor as the sole source of knowledge, and the longer they worked in a cooperative-learning mode the more pronounced this change became. Also noted was that even though a student did not know how to solve a given problem, he or she was more likely to attempt a solution.

2.2.3 Socio-academic considerations

Current studies in cooperative learning seem to indicate that one cannot easily separate the social aspect from the educational aspect. Vidakovic (1992) observes:

> While working together in small groups, students engage in two types of problem solving. On the one hand they attempt to solve their mathematical problems correctly, and on the other hand they have to solve the problem of working productively together. (p. 647)

Neither the social problems nor the mathematical problems are always easily solved by the students. A laboratory assistant observing groups at work reports the following (Wimbish, 1992):

> Thinking back over my observations of all the groups, I believe that it is a combination of social and academic issues that affect a group's effectiveness. Groups who are socially compatible seem to work fairly well together only if they also have a positive attitude toward the class. Groups which have an academic advantage do not function very well until they become socially compatible. Before they become friends, they work more as individuals than a group. (p. 72)

On the other hand, many students are able to make team participation work for them, as the assistant observes later (op. cit.):

> Mary and Anne got off to a slow start today. They spent most of the first part of the period talking and not working. Mary was writing the definitions down while they were talking. I was surprised because after about thirty minutes they started working seriously on the assignment. They would both read the question and propose a possible answer. Then they would discuss the text until one of them understood it and she would explain it to the other. They both shared equally in the work. Even though they got a late start they accomplished more by working well together than any of the other groups. (p. 73)

Thus we see that careful observations of group interaction show that both the social and the educational facets are interspersed throughout a working session.

2.2.4 Structured conversation

One of the difficulties that students and instructors have mentioned when beginning to work cooperatively to solve problems or to construct concepts is knowing how to begin an analysis. Getting started and overcoming apparently dead-end situations can be major hurdles for the beginner and sometimes even for the expert.

The students, Mary and Anne, in the example cited above, were making use of a technique called "structured conversation." Structured conversation is a form of Pólya's heuristic problem-solving method (Pólya, 1945) blended with Socratic-like questioning.

Pólya (1945) suggests that successful problem-solving involves the following steps:

1. understand the problem,

2. devise a plan,

3. attempt to carry out the plan, and

4. look back over the results.

Pólya's method can be combined with the Socratic method by asking questions that lead to understanding the problem. For example, consider the following problem posed to students who have a finance section to study. The students are asked to construct a pension plan for themselves. They must identify how much and for how many years the pension plan is to pay them after retirement. They must determine the amount needed at retirement to support the periodic payments. Finally, they must determine the amount and frequency of their payments into their accounts before retirement; thus they must have some knowledge of interest rates now and in the future. Having identified these issues in understanding the problem, the students can proceed to map a strategy to set up the pension plan. The other steps of the program might consist of doing the calculations needed to understand what money needs to be deposited and when it needs to be deposited. In addition, the students must check their results against the proposed goal. From the cooperative learning point of view, the students discuss each of the issues as they proceed and develop appropriate language to communicate the results.

A process for discussion and conversational feedback

A well-defined procedure of oral communication while working on a problem in a group can help students to clarify an idea for themselves. Certain rigorously defined steps yield a pattern for an effective problem-solving dialogue. The pattern consists of a series of instructions like Pólya's method with one additional step:

1. understand the problem,

2. devise a plan,

3. carry out the plan,

4. look back, and

5. devise a suitable method for communicating the results.

Researchers such as Davy (1983) refer to this as a "think-aloud" process. Students begin to develop good analytic skills when they are encouraged in their early problem-solving sessions to follow a prescribed form of discussion (Wimbish, 1990). Not only do they learn to ask questions of others in their group, but when they are working alone they ask questions of themselves. Most important, they learn to ask those questions that will help them construct the mathematics necessary to solve problems, not just give answers to particular problems.

Among the most important aspects of discussion used in cooperative learning is the possibility of conversational feedback. Dubinsky (1989a, 1989b) and Dubinsky and Schwingendorf (1990) emphasize the need for the individual learner or problem solver to construct (or to reconstruct) her or his own understanding — to construct the idea in her or his own mind. To accomplish these aims, Dubinsky encouraged intensive use of interactive computer programs. Personal computers, besides helping in direct concept construction, provided immediate feedback to students when used with interactive software. Use of the mathematical programming language ISETL, or an appropriate use of the programming features of a computer algebra system such as Maple V, provides students with a way of generating problem-solving conversations. In this way, the computer itself can be established as a cooperative group member and not just a device to perform calculations or plot graphs.

Thinking about thinking: metacognition

Participation in a cooperative group, with or without technological enhancement, has the added benefit of encouraging students to think about their own problem-solving methods and critical thinking skills. Thinking about thinking, *metacognition*, helps students to become more aware of their own problem-solving methods. Experimental evidence (Wimbish, 1990) suggests that examination of thinking processes does improve analytic skills.

Self-managed learning

There is anecdotal and experimental evidence to support the thesis that many students arrive at college with poor self-management skills. Cooperative learning requires self-management of the participating students. Those students lacking self-management skills or who view the instructor as the only class manager can prevent or inhibit the use of cooperative learning interventions. Skinner and Smith (1992) discuss the limitations of using self-management interventions with students who have deficiencies in academic skills or who have poor attitudes toward interventions requiring self-management. It is therefore important to assist students to develop or to improve self-management skills. On the other hand, some studies (Wimbish, 1990; Schwingendorf, Wimbish, and Hawks-Hoover, 1992; Schwingendorf and Wimbish, 1994) demonstrate that cooperative learning can assist students in strengthening their self-management skills. The longer the students work within a cooperative group environment, the less dependent on the instructor they become. They become more willing to explore problems on their own — particularly to explore new, nonstandard problems. And they become more willing to try to explain their ideas to others.

Constructivism

According to the *constructivist theory of learning*, the learner constructs knowledge and does not passively receive knowledge; students come to know by an adaptive process of organizing and making sense of their experiences rather than by perceiving some external reality. Educational researchers have found that solving problems develops mathematical knowledge. Owens (1992) observed that problems and problem-solving can provide an environment for the construction of knowledge. Problems and problem-solving establish a need for mathematical knowledge and a context in which students can learn mathematics. However, problems generate mathematical knowledge only to the extent that students perceive those problems as their own, find the knowledge thus generated of practical use in solving a range of problems, and perceive a value in expanding a conceptual domain.

Yackel, Cobb, and Wood (1991) explain that from the constructivist view, students do not learn mathematics by internalizing carefully prepared and completely organized concepts and procedures. They learn mathematics by doing the organizing or the reorganizing activities for themselves. These activities include both *thinking* and *being aware of one's own thinking process*. These two types of thinking occur while conversing with oneself or with others. Cobb and Steffe (1993) note that constructivism in the classroom can take the following form: the instructor provides a set of activities, and the students work on the activities. Work on the activities includes trying to establish patterns of mathematical ideas and trying to reorganize existing patterns into new patterns.

In the present context, mathematical activity consists of thinking and idea constructing. Mathematical thoughts and ideas can be constructed by individuals acting alone as well as by groups of individuals acting together. Students can be guided to construct mathematical ideas for themselves by working cooperatively with other students and with the instructor. Dubinsky (1989a, 1989b) and Dubinsky and Schwingendorf (1990) emphasize the need for individual learners or problem-solvers to construct or to reconstruct their own mathematical knowledge in order to activate the learning process.

Balacheff (1990) merges constructivism and learning theory to explain another approach to mathematical learning. He characterizes the learning and teaching process in mathematics as a relationship between two theories about how students learn, constructivism and learning through problem-solving, and two constraints, the nature of mathematical knowledge and the nature of the classroom. Balacheff emphasizes that mathematical knowledge is constructed by the learner or doer of mathematics through solving problems. Dubinsky observes that, in a very real sense, problem-solving fosters concept construction and that students' progress in mathematics can be measured by the variety and difficulty of problems that they can solve.

Is it necessary or even helpful for students to work in groups? Vidakovic (1993) reports that students who worked together in cooperative learning groups were as able to construct mental concepts appropriate for calculus as the individuals working alone; however, students who worked in groups were able to construct these concepts more quickly and more thoroughly.

2.3 Piaget's theory of cognitive development

Piaget's view is that individuals construct knowledge by interacting with their environment. Even though social interaction, which is a part of this environment, may enhance an individual's development, it cannot change the course of this development in essential ways.

According to Piaget (Piaget & Inhelder, 1969a), a person's intellectual development is influenced by maturation, experience, social interaction, and equilibration. We will briefly discuss each of these factors.

Piaget's work focused primarily on young children. He observed that a child's physical, social and emotional maturation were intrinsically related to the child's intellectual development. The construction of various kinds of mental images is possible at specific stages in a person's overall development. *Maturation* determines whether or not the construction of specific mental structures is possible at a particular time.

Piaget distinguishes two types of experience at elementary levels: physical experience and logico-mathematical experience. As the child grows older, he or she becomes capable of deductive thinking and conscious abstraction or mental construction (called *reflective abstraction*), both of which are based on these two kinds of experience.

Physical experience consists of acting upon objects, that is, playing around with things. Children play with things by transforming them and interacting with them using all their senses — touch, sight, taste, smell, and sound. For children, play is their way of thinking. Children draw empirical knowledge about the objects they are playing with — color, size, shape, texture, and so on — by reflecting on the results of their actions.

Not all knowledge is drawn directly from simple actions on the objects themselves. For example, if a child places 10 marbles in a row and counts them, then rearranges them into a circle and recounts, then into a rectangle and recounts again, it is a collection of separate physical experiences, each of which has a numerical result. But when the child reflects on these various actions and coordinates them in the realization that, not only was the answer always the same, but no matter how the marbles are arranged, the count will always be 10, then this is knowledge gained from *logico-mathematical experience*.

Thus, whereas physical knowledge comes from actions on objects, logico-mathematical knowledge has as its source the general coordination of these actions.

Human beings learn about an idea from experience and not merely because someone tells them about it. Piaget emphasized that *social interaction and transmission* is insufficient by itself. In fact, in the process of social interaction, the individual contributes as much as he or she receives. Even in interactions where an individual appears most passive, there is no social action without active assimilation. Thus, when a student is nodding off to sleep in the back of the classroom — even if the instructor is giving a dynamic lecture — there is no social interaction, and thus no transmission of knowledge. On the other hand, in a small group discussion, even if one student is doing most of the talking, the other group members who may be listening passively are engaged in a social interaction. In the following section we describe this factor in more detail.

When an individual is presented with situations that require new mental constructions, the tendency is to try to fit these new ideas into existing mental structures, that is, to *assimilate* this new knowledge. If this doesn't work, the individual experiences disequilibrium. Changes must be made to *accommodate* these new experiences. *Equilibration* is the internal mechanism that regulates the processes of assimilation and accommodation.

2.4   Social interaction and learning

Although Piaget did not write extensively on the topic of social interaction, his work contains a number of implications concerning the role of this kind of experience, particularly the importance of interactions among peers in the learning process. In his studies of children's development of speech, moral judgment, logic and language, he emphasizes that communication with others enhances the child's ability to see things from another's point of view. In talking about problems, children develop the ability to pay attention to more than one element of a problem at a time. (Piaget, 1926, 1932; Piaget & Inhelder, 1969a, 1969b).

Social interactions with other individuals also involved in the learning process provide opportunities to see that others may arrive at different conclusions. These differences may give rise to conflict among the individuals in the group. Contradictions coming from others at a similar level of conceptual development serve to bring the cognitive differences into sharp focus, and thus lead to coordination that can resolve the conflict.

Piaget suggested that opportunities for becoming more able to see another's viewpoint are much more common when learners discuss things with one another (Piaget, 1968, 1970a). Individuals who wish to communicate and to be understood must adapt to the informational needs of the listener. Each person must face the reality of different human perspectives when involved in active group discussion. Cooperation with other students in the discussion helps the student to learn how to take different points of view into account. When students share a goal, the result of trying to reach it can, because of different perspectives, lead to cognitive conflict. Resolving such conflicts leads directly to cognitive development.

Another situation in which Piaget clearly acknowledged the role of social interaction was that of *operational grouping*. (Operational grouping is a model used to explain the mental processes involved in the formation of concepts.) He emphasized that individuals would never come to group operations into a coherent whole without interchange of thought and cooperation with others. Piaget put it in the following way:

> The grouping consists essentially in a freeing of the individual's perceptions and spontaneous intuitions from the egocentric viewpoint, in order to construct a system of relations such that one can pass from one term or relation to another belonging to any viewpoint. The grouping is therefore by its very nature a coordination of viewpoints and, in effect, that means a coordination between observers, and therefore a form of cooperation between several individuals. (Piaget, 1950, p. 164).

Thus we see that Piaget's ideas of how individuals — adults as well as children — go about the process of constructing knowledge supports the practice of having students work together in cooperative learning groups.

2.5 Chapter Summary

Research and anecdotal evidence suggest that working cooperatively to learn mathematics affects students' attitudes and beliefs about what mathematics is, how mathematics is created, and whether "ordinary people" — that is, the individual students themselves — can be expected to learn mathematics. When beginning to work cooperatively, students (and their instructors) are actually learning both to construct mathematical ideas and to cooperate with others in this process. There is an active interplay between the construction of mathematical concepts and the development of cooperative social skills.

Structured conversation, the use of clearly defined steps for discussion and conversational feedback, helps to create an environment in which students learn to ask questions of others — and of themselves — as a part of the process of learning mathematical concepts. Problems and problem-solving provide a context that create a need for mathematical knowledge. Participation in cooperative groups tends to create situations in which students reflect on their problem-solving processes, and this reflection helps to strengthen students' analytic skills. Students in groups can develop the skills necessary for solving a wide variety of problems, both social and academic.

Working cooperatively in small groups provides students with opportunities to interact socially with their peers in solving problems. In attempting to resolve the conflict that arises when group members find different "answers" for the same problem, the students engage actively in processes that lead directly to cognitive development.

In any decision to use cooperative learning groups as a principal mode of instruction, instructors should be aware that not everyone means the same thing by the term. The authors advocate a model of cooperative learning based on the guidelines given in Chapter 1. These guidelines were formulated from well-established principles of learning theory, and have yielded positive results. By working cooperatively in small groups, students can develop skills necessary for solving a wide variety of problems, both social and academic. And in solving these problems, they do construct robust mental images of the mathematical concepts that they are studying.

Chapter 3

How Are Groups Formed?

In this chapter we describe some basic ideas on the organization and composition
of a typical group: the size of a group; the distribution of talent, expertise, social
characteristics and student preferences in groups; some ways to form groups; and
the lifetime of a group. The recommendations in this chapter are based on the
experiences of the authors of this monograph, the respondents to our survey,
and some other colleagues, as well as recommendations in the existing literature
on cooperative learning (for example, see Davidson (1989a,1990b), Johnson and
Johnson (1984, 1985, 1987a, 1990), and Johnson et al. (1984)).

3.1 The size of a group

The authors recommend that instructors strongly consider forming groups that
are heterogeneous in nature (as described in Section 3.2) and that each group
contain three or four students. Heterogeneous groups are more likely to bring
together a mixture of life experiences that give rise to multiple viewpoints as
students work to solve problems. We find that the size of a group affects its
ability to be productive. There is consensus in the literature on cooperative
learning that the ideal size of a group is four students. What are some of the
possible reasons? Four students can split into subgroups of two students each
and then report back to the group; pairs of students can easily work at one
computer; pairs can work on skill-oriented drill activities; heterogeneous groups
of four permit an adequate combination of individual talents and resources as
well as the possibility of gender balance (with two females and two males); a
group of four can sustain itself if one student is absent or drops out of the group;
a group of four encourages more effective work habits, structured conversation,
and reflective thinking.

Since students' structured conversation fosters the construction of their own mathematical understanding (Section 2.2.4), groups should be amenable to such conversation. Discussion in a two-member group more frequently falters than in a group of three or four members. On the other hand, individuals in a group of five or more students may not have enough "air time" during discussions. The authors find that each student in a group of three or four has the opportunity to participate and that, in these groups, there usually is a sufficient variety of ideas and expertise to sustain the conversation.

In addition, because good individual self-management skills enhance the possibility of effective structured conversations within the groups, the authors generally avoid forming groups of more than four students. A large group may find that it is difficult to get organized, manage group cooperation, coordinate work, and reach agreement. A student in a smaller group, where the group's organization is more workable, has a better chance of developing individual management skills in concert with the other group members.

In the computer laboratory, groups of three or four students can work well together at two adjacent microcomputers, while a group of five may not be able to gather around two computers comfortably. Conversation involving the entire group is possible only when the group members can work near to each other. When members of a group can use two adjacent computers in the laboratory, they often get one problem going on one machine, and then they begin work on a second problem on the adjacent machine. At those times when two problems are related, the discussion can be quite fruitful.

There may be circumstances where a two-member group is necessary or useful, such as on a commuter campus, especially where students commute long distances. We discuss commuters later in this chapter.

Although the literature doesn't offer much support for groups of three, the authors of this monograph have found them to be quite effective at times. The authors and the respondents to our Survey on Cooperative Learning have observed that a group of four may subdivide into two separate groups of two that do not report back to the whole group, while a group of three generally remains a group of three. Groups of three are, however, more greatly affected by the absence of any one of the group members. Although in particular groups members may not work well together (as described in Chapter 6), some of the authors have found groups of three members to be effective about as often as groups with four members.

The authors recommend groups of three or four students in most circumstances. In many classes, we initially arrange groups of three and four with as many groups of four as possible. For example, in a class of 33 students, we would try to form six four-member groups and three three-member groups — rather than eleven groups of three members. Later, when students drop or add the class, more groups of three, or even groups of five, become necessary.

3.2 The composition of the groups

The authors recommend that instructors distribute the talent, expertise, and various social characteristics represented in the class to form heterogeneous groups. While the groups have to be socially compatible and able to work together outside of scheduled classes, we have observed (and research into how individuals learn mathematics supports our observations) that conversations about problems are richer and more productive if the groups represent as broad a base of life experiences as possible (see Section 2.4). We use a significant number of the following criteria in forming groups:

- major or area of study interest;

- geography — including address (Other information — such as phone number(s) and e-mail address — may be requested here although these are not criteria for group formation.);

- social characteristics, e.g., age, gender, race, nationality and/or ethnicity, and, depending on the situation at the college, whether or not students prefer being in a group with students of their own gender, race, nationality and/or ethnicity;

- weekly time schedule including times of classes and labs, as well as any work, family, or other responsibilities;

- academic background, e.g., mathematical expertise, including strength of mathematics background, grades (success) in mathematics courses, and mathematics SAT or ACT scores;

- verbal SAT or ACT scores;

- previous cooperative learning group experience and the type of experience;

- computer and/or calculator experience;

- keyboarding or typing experience; and

- student preferences for any other individuals to be in the same group.

Student preferences may be determined after a week or so of introductory lessons on material with which they are comfortable or that is not too difficult. Such an introductory period can be used to guide students in working cooperatively and at the same time allow students who don't know others in the class to get to know some of their classmates.

Some of the authors and some of the respondents to our survey have collected information for group formation by using a student questionnaire and a time schedule sheet like those in Appendix C. The students fill out these forms early in the semester. Some of us have found it useful to wait until the end of the first week or early in the second week of classes to have students fill out the time schedule sheet to avoid problems or confusion caused by subsequent changes in student class and work schedules. We discuss further the timing and process of group formation in the next section of this chapter.

We should note that when there is a broad range of ability (for example, in SAT or ACT scores) in a class, it is usually better not to place the very best students with the weakest in the class. Some of us have found it quite useful to place the top quarter of the students together, the bottom quarter together, and the middle quarters together, respectively.

There are some who suggest that the importance of geography decreases with increased access to phones, e-mail, fax, and so on. However, some of the research literature on cooperative learning (Johnson and Johnson, 1984) suggests that face-to-face interaction is of significant importance if cooperative learning is to be as effective as possible. While technology indeed makes communication easier, the important team-building process of group members being able to get together and interact face-to-face can be best achieved if they have ample opportunity to meet as a group. Hence, it is important to take into account the class, work, personal and outside activity schedules of the students when forming groups. Especially on a commuter campus, taking into account the geographical data of the students will promote fact-to-face student interaction in groups outside of the classroom and laboratory environments. In fact, students who have had two or three semesters of successful work in classes that have used cooperative learning groups have asserted that — from their perspective — the single most important criteria for forming a successful group is the ability to meet with their group members on a regular basis outside of class.

Each instructor's personal situation and experiences with the use of cooperative groups is a guide in deciding what information about students will be most important in the group-formation process in a particular class. The authors' experiences indicate that the formation of heterogeneous groups in the college mathematics classroom can be very useful in meeting the objectives of team building and the development of an *esprit de corps* within and among groups.

3.3   The process of group formation

We begin with a description of one method of group formation that some of the authors and respondents to the survey have used to form heterogeneous groups in a first semester calculus class with a computer laboratory component. As mentioned earlier, Appendix C includes sample questionnaires and a sample time-schedule form that have been used to collect data for the process of selecting heterogeneous groups. Once the questionnaire and/or time-schedule forms have been completed, the instructor selects several primary criteria for group formation; for example:

- computer expertise and typing skills (if computers are used in the class),

- mathematics background and expertise (including SAT or ACT scores and previous mathematics courses),

- SAT or ACT verbal scores, and

- gender.

The remaining data on the student information sheets (mentioned in the previous section on characteristics used in forming groups) are then used as secondary criteria in group selection. On a commuter campus geographical considerations are added to the list of primary criteria. In subsequent semesters of a course sequence, a shortened questionnaire can be used along with the time schedule sheet. Student preferences for other individuals with whom to work (and *not* work) should be given consideration, especially in subsequent courses.

Some of the authors have found it helpful to have two or more instructors work together in the group selection process, particularly if these instructors are teaching the same course or sequence of courses. In any case, we begin by using the primary criteria for selecting the first one or two students to form a foundation for each group. Then, on a second pass, one or two individuals can be added to each group. Finally, using the secondary criteria and remaining data on students, the group selection can be completed.

The authors strongly recommend that groups be formed as soon as possible in the semester so that the process of building commitment to a group and the building of an *esprit de corps* can begin promptly. However, some instructors take as long as two weeks to form final permanent groups. One way to do this is to form pairs, and then — after observing student work habits in problem-solving and/or computer-laboratory situations — to pair the pairs or make other adjustments to form the final groups. However, instructors should be careful not to take too much time to complete the selection process. Such a process can easily be integrated into the computer laboratory and classroom activities. Once groups are formed, their first assignment — a first-step in the team-building process — is to choose a "team or group name." Whether done right away or during the first weeks, the authors believe it is best to have permanent groups established by the end of the second week of the class.

3.3.1 Involving students in the group formation process

Although on campuses with a large proportion of traditional-aged resident students the instructor may take the primary role in the formation of groups, there are circumstances where student participation in formulating the process of group formation may be desirable. Some of the authors have had much success by selecting groups based on student preferences. A commuter campus presents a special and common situation in which there may be more older adult students for whom geography and time schedules are the most important considerations in group formation. In such a situation, and possibly others, the involvement of students in the decision of how groups are formed may be a very useful and powerful consideration. In general, students will need to "buy into" the idea of cooperative learning. In particular, adult students often express a desire to have a feeling of control over their lives. Inviting their suggestions on the process of group formation may help to strengthen their commitment to and identification with their own group. Instructors may have students contribute to the list of criteria that they feel are important in forming groups. If the students' list omits critical criteria, the instructor then suggests or prompts students to include these missing criteria and discusses their merits.

There are various ways to involve students in the process of group formation. During the first two weeks of class, the students may engage in a variety of class and computer-lab activities in different informally-chosen small groups. Toward the end of the first week or the beginning of the second week, students can participate in a full-class discussion in which they introduce themselves and speak about the strengths they can bring to a group. The instructor can also invite each student to speak about what he or she will need and hope to receive from other group members. Involving students in such a process might lead to a very sensitive discussion. The instructor needs to take an active role in facilitating such a discussion and ensuring that there are no breakdowns based on individuals' lack of self-esteem. This kind of discussion seems to be more successful and productive after the students have had a little time to get to know each other. Such a discussion may even be useful before students fill out a questionnaire.

Among the authors and the respondents to our survey, there was a clear consensus that forming groups by a pseudo-random process, such as alphabetically or by student seating preferences, may not be conducive to developing an effective cooperative learning environment. Such a process may, at best, serve as a first approximation for the college classroom.

3.3.2   Student self-selection into groups

Some of the authors and some of the respondents to our survey have reported success with involving students in a self-selection process for group formation. In the first class period, the students may be given a problem and asked to work in informal groups toward solving the problem. This activity is followed by a discussion on "What makes a group function well?" After at least a week of working in various informal groups on problem-solving tasks, the students are invited to introduce themselves, to offer some observations about what makes a group function well, and to mention some strength that they feel they will be able to bring to a group. Finally, toward the end of the second week of the course, the students are allowed to select their groups. The conclusion of this activity is to name their group. One instructor asks the groups to make some kind of statement about what criteria they used in selecting their groupmates and how these criteria are met in their particular group of students.

3.4 The lifetime of a group

In the cooperative learning literature, groups that have a lifetime of one or more semesters are sometimes referred to as *cooperative base groups*. The authors believe that to achieve the kind of *esprit de corps*, commitment to group learning and team building needed for an effective cooperative learning environment, the use of cooperative base groups provides a milieu amenable to achieving the best possible results. Some of us have had experiences where students made strong requests to stay together. Such situations argue against an instructor being too inflexible. In the case of a crisis, such as when all but one student from a particular group drop the course, or two students drop from each of two different groups, or if there is a seriously dysfunctional group, then the instructor reorganizes some groups. In such crisis situations when group rearrangement is necessary, the authors strongly recommend involving at least the students of the groups to be adjusted, and possibly involving the entire class in the rearrangement considerations. We say more about this in Chapter 6.

3.5 Chapter Summary

In this chapter the authors respond to the question "What is a group like?" by describing some basic ideas on the organization and composition of a typical group: the size of a group; the distribution of talent, expertise, social characteristics and student preferences in groups; some ways to form groups; and the lifetime of a group.

There is a consensus among the authors, which is also reflected among the respondents to our survey, that groups be formed with three or four students, and that each group contain a heterogeneous mixture of some of the various academic and social characteristics represented within the class. Specifically, the authors and the survey respondents recommend that groups represent a heterogeneous mixture of majors, social characteristics, academic background, prior experience in cooperative learning groups, and familiarity or experience with the various forms of technology that will be used in the course.

Although we recommend mixed abilities within a group, we have found it best not to put the very strongest students with the very weakest. Some of us have obtained good results by mixing students within the quartiles — putting students from the top quartile together, those in the middle quartiles together, and students from the lowest quartile together.

Students who have accepted the whole cooperative learning process report that it is necessary for groups to be able to meet outside of class; thus schedules and geography are important criteria in group formation. It is helpful to consider student preferences, especially in subsequent courses. Some of our students have made strong requests to stay together in the same group from one semester to the next.

Some instructors assume a primary role in group formation, others report good success with involving students in the process, or even guiding the students in a self-selection process of group formation.

There is strong consensus among the authors and the survey respondents on forming the groups early in the semester — perhaps toward the end of the second week of classes — after giving the students several opportunities to work in various informal groups.

The literature suggests that *cooperative base groups* with a lifetime of one or more semesters provide a social context conducive to achieving the kind of *esprit de corps* in which commitment to group learning flourishes. The experience of the authors is that such groups help to create and sustain an effective cooperative learning environment.

Chapter 4

What Do Groups Do?

Students working in cooperative learning groups *do mathematics* as they learn mathematics. Now that we have discussed the formation of groups (in Chapter 3), we will examine just what the authors ask their students to do in groups that helps the group members to construct their own knowledge of mathematical ideas. In particular, we will describe experiences that promote learning through group interaction during scheduled times in the classroom and computer laboratory as well as during group homework sessions.

The key idea here is that students can be guided through a three-step process as they develop their own understanding of a mathematical concept. The cooperative learning groups tackle different kinds of problems at each of the three steps:

1. informal *activities* that introduce a concept, frequently through the investigation of examples;

2. classroom *tasks* that pull together the germs of ideas developed in the activities; and

3. *exercises*, including traditional homework problems, that require application or extension of the concept, and that reinforce the concept being studied.

The traditional lecture method of teaching includes some aspects of these three steps: presentation of some examples as motivation; formal statement of the definitions, theorems, and proofs in the same lecture; assignment of some reinforcing exercises, probably attempted by students in isolation. However, the instructor using the lecture method is the only person required to be active in the first two steps, and a student may reinforce ideas only as far as he or she can progress alone. Cooperative learning groups in the authors' classes actively participate in each of the three steps. Since the preliminary activities are performed before class presentations, the students are already engaged in thinking about a concept before it comes up in class. The class time is not spent in presentation of an "oral text" since we assume that the students' (written) text book formally presents the mathematics in an organized fashion. The students interact with the instructor and each other rather than simply accepting information, and at each step the students can confer with others in their group as they actively work on problems. By listening to the thinking-aloud processes of the other students in their group as they work together to develop an understanding of a concept, many students are led to think about and comprehend better their own thinking processes (see Section 2.2.4). Talking about problems does help the students to stay aware of more than one aspect of the problem at a time (see Section 2.3).

In this chapter, we will consider each part of this process separately and give particular examples of mathematical problems that the students would be working on at each stage. Then we will discuss what students in cooperative learning groups do with computers, and what they do in a computer laboratory. We address group examinations in Chapter 5.

4.1   Activities before a concept is studied

We use the word *activity* to refer to a mathematical problem that foreshadows the introduction of a concept. These activities are carefully designed problems that lead the students to understand, to make conjectures, or to believe a mathematical idea as they think deeply about the related concept. An activity may involve observation of a pattern, construction of a model, or translation to mathematical notation. We recommend that groups work on the activities during scheduled laboratory or problem sessions as well as during group meetings that the students arrange among themselves.

Research in mathematics education suggests that students learn mathematics by constructing mathematical ideas in their own minds (see Section 2.3). They learn by doing mathematics and by being active rather than passive learners. The activities attempted by cooperative learning groups engage the students in structured conversations and encourage individual group members to confront mathematical ideas before formally studying about them (see Section 2.2.4).

The purpose of an activity is to create a basis for class discussion of a concept, rather than to find unique, numerically-correct solutions. Some of the authors and respondents to our survey have found that the work students do on the activities before formal instruction is valuable even if the students do not arrive at "correct answers." The group activities offer a framework for the students to engage in structured conversations. By grappling with an issue during an activity, the students become interested in finding a solution for the problem. And they begin to reorganize their thoughts and construct their own mental concepts.

Computers and graphing calculators can provide students with a kind of "conversational feedback" as they work on these preliminary activities (see Section 2.2.4). In this case, the activities are computer-based problems that require students to use a symbolic computer system or write short code segments in a mathematical programming language to construct mathematical ideas. The work that students do with their group members in the computer laboratory is discussed in Section 4.4.

4.1.1   Examples of activities

Examples of activities suitable for various courses follow. For each example, we state the related concept and briefly describe the activities. Students may need a fuller explanation of the activities.

- **Activity Example 1**

  Course: Precalculus, Trigonometry

  Concept: Identities involving sine and cosine functions

  Activity: 1. Graph (using a graphing calculator or computer) the functions f and g defined by:

  $$f(x) = \sin^2(x) \quad \text{and} \quad g(x) = \cos^2(x).$$

  Then graph the function $f + g$. What do you observe?

  2. Graph the cosine function. Then graph the function f given by:

  $$f(x) = \sin(x + c)$$

  for various values of c until you find some values of c that make the graph of f appear to coincide with the cosine graph. Be sure to check that the graphs coincide over several different intervals. List several values of c that work and express these as approximations involving the number π.

  3. Do the preceding problem with the roles of the sine and cosine functions interchanged.

  4. Graph the functions f and g defined by:

  $$f(x) = \cos(2x) \quad \text{and} \quad g(x) = \sin^2(x).$$

How can the graph of g be transformed (by shifting, stretching, and/or reflecting about a line) so that it coincides with the graph of f? How can the function f be expressed in terms of the function g?

- **Activity Example 2**

Course: Calculus I

Concept: Properties of exponential functions and the number e

Activity: Consider the exponential function:

$$f(x) = b^x.$$

1. Find a number b_1 between 2 and 3 such that the graphs of f and its derivative coincide. Estimate the number b_1 as accurately as possible by zooming in.

2. Set up a difference quotient for the function f:

$$dq(x) = \frac{f(x+h) - f(x)}{(x+h) - x}$$

and simplify as much as possible.

3. Find a number $b_2 > 1$ so that:

$$\lim_{h \to 0} \frac{(b_2)^h - 1}{h}$$

is very close to 1.

4. Compare the numbers b_1 and b_2. Do you recognize this number?

- **Activity Example 3**

Course: Mathematical Modeling, Numerical Analysis

Concept: Interpolation and curve fitting

Activity: Consider the points $(-1, 1)$, $(1, 2)$ and $(3, -1)$.

1. Find a polynomial of degree 2 whose graph passes through these three points, and graph the polynomial. Is this the only polynomial of degree 2 that works? Why or why not?

2. Find a polynomial of degree 3 whose graph passes through these same three points, and graph the polynomial. Is this the only polynomial of degree 3 that works? Why or why not?

3. Find a polynomial of degree 1 (a linear equation) whose graph passes through these three points, and graph the polynomial. Is this the only polynomial of degree 1 that works? Why or why not?

(Some students expect that every question posed in a mathematics class — by either the instructor or the textbook — has a "right answer." The fact that there is no polynomial of degree 1 that passes through these three points is disturbing to some students and can create a temporary disequilibrium in their minds. A significant shift takes place in the mind of a student who begins to question whether there is a solution, and who begins to try to explain why there is no solution.)

- **Activity Example 4**

Course: Linear Algebra

Concept: Inverse matrices

Activity: *(These activities are to be done in a computer laboratory using appropriate software.)*

1. Given the following matrices:

$$A = \begin{pmatrix} 1 & -2 & 5 \\ 3 & -7 & 0 \\ -4 & 0 & 6 \end{pmatrix}, \quad B = \begin{pmatrix} 2 & 4 & -1 \\ 0 & 3 & 7 \\ 1 & 5 & 9 \end{pmatrix}, \quad C = \begin{pmatrix} -3 & 0 & 9 \\ 2 & 6 & 1 \\ -7 & 3 & 6 \end{pmatrix}$$

Find each of the following matrices:
   (a) A^{-1}, $(A^{-1})^{-1}$, B^{-1}, $(B^{-1})^{-1}$
   (b) $(6A)^{-1}$, $\frac{1}{6}A^{-1}$, $(\frac{2}{3}B)^{-1}$, $\frac{3}{2}B^{-1}$
   (c) $(AB)^{-1}$, $A^{-1}B^{-1}$, $B^{-1}A^{-1}$, $(BC)^{-1}$, $C^{-1}B^{-1}$, $(AC)^{-1}$, $C^{-1}A^{-1}$
   (d) A^6, $(A^6)^{-1}$, $(A^{-1})^6$
   (e) $(A^t)^{-1}$, $(A^{-1})^t$, $(B^t)^{-1}$, $(B^{-1})^t$,

2. Make some conjectures based on your computations in part 1.

3. Let X and Y be invertible matrices, and k be a nonzero real number. Evaluate or simplify each of the following expressions.
   (a) $(X^{-1})^{-1}$
   (b) $(kX)^{-1}$
   (c) $(XY)^{-1}$
   (d) $(X^k)^{-1}$
   (e) $(X^t)^{-1}$

- **Activity Example 5**

  Course: General Statistics

  Concept: Least squares regression line

  Activity: 1. *(Students are given a scatter plot of points for which a straight line seems like a reasonable model. Several straight lines are drawn on the diagram.)* Choose the line that best seems to fit the data. Describe why you chose this line.

     2. *(Students are given a similar plot with no lines drawn.)* What rules might you use to determine a line that fits the data well? State at least two such rules.

- **Activity Example 6**

  Course: Finite Mathematics, Mathematical Modeling

  Concept: Linear programming, graphical method

  Activity: *Power Machines, Inc.* is a small manufacturing firm that produces motorcycles and snowmobiles. Two machines are used in the production of both of these vehicles. Producing a motorcycle requires two hours on machine A and two hours on machine B; producing a snowmobile requires one hour on machine A and four hours on machine B. Machine A can be operated up to ten hours per day, while machine B can be operated up to sixteen hours per day. Each motorcycle that is sold brings a profit of $500, and each snowmobile brings a profit of $700. Assuming that all the motorcycles and snowmobiles that can be produced will be sold, how many of each should be manufactured to maximize the company's profit?

     1. The plant manager at *Power Machines, Inc.* has to decide how many of each type of vehicle to produce. Let x represent the number of motorcycles, and y the number of snowmobiles to be produced. These are called the *decision variables*. Translate the problem into a system of linear inequalities that represent the *constraints* of the problem situation.

     2. Sketch the feasible region (i.e., the solution set) of this system of inequalities, and determine the coordinates of its vertices.

     3. The goal, or *objective*, is to maximize profit. Find a function, $P(x, y)$, that represents the profit realized if x motorcycles and y snowmobiles are produced and sold.

     4. Evaluate the objective function P at some points in the feasible region, at some points along the edge of the feasible region, and at the vertices of the feasible region.

5. What is the maximum value of P? How many motorcycles and how many snowmobiles should be produced to realize this maximum profit?

(The students should be instructed to bring any graphs they sketch to the next class for discussion.)

- **Activity Example 7**

   Course: Intermediate Algebra

   Concept: Quadratic models

   Activity: 1. Use your graphing system to graph the following quadratic functions on one set of coordinate axes:
   $$f(x) = -x^2 + 6x$$
   $$g(x) = -x^2 + 6x + 10$$
   $$h(x) = -x^2 + 6x - 5$$
   (a) Note any similarities or differences among these graphs.
   (b) From the graphs, find the x-intercepts of each function.
   (c) Find the x-coordinate of the vertex of each parabola.
   (d) Is there any relationship between the x-intercepts and the x-coordinate of the vertex of each parabola?

   2. The market analyst at *Vibrant Videos* has found that the profit P in dollars that can be realized by the sale of video tapes is a function of the selling price x, in dollars, of each tape; profit is given by
   $$P(x) = -x^2 + 29x.$$

   Find the maximum profit, and the price per tape that gives this maximum profit.

   3. Laura is planning her garden, and has 180 meters of fencing material. What are the dimensions of the largest rectangular region that she can enclose with this much fencing material?

- **Activity Example 8**

   Course: Linear Algebra

   Concepts: Linear combinations, linear independence, spanning set, basis of a vector space

   Activity: Use your symbolic computer system as you work on these activities.

   1. Let $v_1 = [1, 2, 1, 2]$, $v_2 = [-2, 1, 0, 3]$, $v_3 = [0, -1, 2, 0]$, and $v_4 = [3, 1, 4, 1]$ be vectors in the vector space $\mathcal{R}^4$.

A vector, v, in $\mathcal{R}^n$ is a *linear combination* of $v_1, \ldots, v_n$ if there exist scalars $c_1, \ldots, c_n$ (not all zero), such that

$$v = c_1 v_1 + \cdots + c_n v_n.$$

Vectors $v_1, \ldots, v_n$ are *linearly independent* if

$$(c_1 v_1 + \cdots + c_n v_n = 0) \quad \Rightarrow \quad (c_1 = 0, \ldots, c_n = 0)$$

  (a) Show that every vector $[x, y, z, w]$ in $\mathcal{R}^4$ can be expressed as a linear combination of v_1, v_2, v_3, and v_4.

  (b) Show that the vectors v_1, v_2, v_3, and v_4 are linearly independent.

 2. Show that the set of vectors:

$$[1, 2, 1, 2], \quad [-2, 1, 0, 3], \quad [0, -1, 2, 0], \quad [3, 1, 4, 1], \quad [2, 0, 2, 1]$$

are *linearly dependent.*

 3. Can every vector in $\mathcal{R}^4$ can be expressed as a linear combination of v_1, v_2, and v_3 (given in part 1 above)? Why, or why not?

(It is assumed that the students are using a symbolic computer system or graphing calculator to solve the systems of linear equations that are required to do these proofs; they are not expected to do all of these calculations by hand.)

4.2   Tasks in the classroom

After the students have attempted the *activities* during scheduled laboratory or problem sessions, the authors recommend that most of the following class period(s) be devoted to *tasks*, problems that students work on in their groups and that then are discussed by the whole class. The tasks usually can be completed in less than five minutes. They are designed to promote conceptual understanding and to help students develop the related mental images in their own minds. Students who have worked on the activities, the problems that presage the formal study of concepts, come to class with an experiential basis and intuitive familiarity with certain mathematical ideas. These students are prepared for class discussion, and they can relate meaningfully to a more rigorous study of the concepts.

The instructor poses questions and guides the discussion. He or she may introduce a new concept in an informal and intuitive way, but there is little if any traditional lecturing. There is no attempt to cover all the material (since it is assumed that the students' text develops all the necessary mathematics in an organized and thorough way).

In class, members of each group sit together and act as a team. All groups work on the same class tasks, and each group develops its own plan to work on the given problems. Students reflect on their own group's experience and listen to strategies and solutions tried by other groups. The authors have observed variations on the following procedure for effective group problem-solving in class:

> Individuals read the problem, and formulate an answer or think about a method to be used. The classroom is very quiet as the students reflect on the problem. After a short period of silence, students have thoughts or questions for members of their groups. The group members explain to each other their ideas and suggested methods to solve the task. Using structured conversation (see Section 2.2.4) the individual group members offer their ideas and receive conversational feedback from each other. The group decides on a strategy to solve the problem, carries out this strategy, and formulates a group answer to the given problem.

During the group work, the instructor may move from one group to another to observe their progress and to provide assistance by giving hints, asking students to clarify a point, discussing a problem-solving strategy, clarifying some notation, and asking or answering some questions. With such a procedure, each student has time to understand the problem and to discuss it comfortably with a few students that he or she knows well.

When several groups seem to have reached a group solution or when a predetermined time has passed, the instructor asks one of the groups to present the group answer (or partial answer) as a basis for discussion for the whole class. Other groups may extend, accept, and criticize the response as the discussion continues. At the conclusion of the discussion, the instructor may tie together some ideas, introduce formal definitions and theorems, or affirm effective problem-solving strategies.

The spokesperson for a group during the discussion may be chosen in various ways. At the beginning of each class period, each group may choose a group representative to speak for the group. Or different members of the group may be encouraged to speak during any class period. The instructor may call on an individual student to present her or his group's ideas to the class. Over the course of the semester (if not in every class period), each member of the class should have the opportunity to represent the group and interact with the instructor during the whole class discussion. In this way, individual student involvement is maximized.

4.2.1   Examples of tasks

Here are some examples of class tasks related to the activities of Section 4.1.1.

- **Task Example 1**

  Course: Precalculus, Trigonometry

  Concept: Identities involving sine and cosine functions

  Task: 1. How do the definitions of the circular functions tell us that

  $$\sin^2(x) + \cos^2(x) = 1?$$

  For which values of x does this equation hold? What is an identity?

  2. What is the smallest left horizontal shift of the sine graph that will move it on top of the cosine graph? Does this give us an identity? What are some other similar identities?

  3. State an identity relating $\sin^2(x)$ and $\cos(2x)$. How does this follow from an identity involving $\cos(x + y)$?

- **Task Example 2**

  Course: Calculus I

  Concept: Properties of exponential functions and the number e

  Task: The number you were looking for *(in the activities you called it b_1 or b_2)* is called e. What is the derivative of the function f defined by

  $$f(x) = e^x?$$

  Use the definition of derivative to find $f'(x)$.

- **Task Example 3**

  Course: Mathematical Modeling, Numerical Analysis

  Concept: Interpolation and curve fitting

  Task: Suppose that you have the following set of data points:

  $$\{(x_0, y_0), (x_1, y_1), (x_2, y_2), (x_3, y_3)\}.$$

1. Does the following expression represent a polynomial that goes through each of these points? How do you know?

$$P(x) = \frac{(x - x_1)(x - x_2)(x - x_3)}{(x_0 - x_1)(x_0 - x_2)(x_0 - x_3)}y_0 \quad +$$

$$\frac{(x - x_0)(x - x_2)(x - x_3)}{(x_1 - x_0)(x_1 - x_2)(x_1 - x_3)}y_1 \quad +$$

$$\frac{(x - x_0)(x - x_1)(x - x_3)}{(x_2 - x_0)(x_2 - x_1)(x_2 - x_3)}y_2 \quad +$$

$$\frac{(x - x_0)(x - x_1)(x - x_2)}{(x_3 - x_0)(x_3 - x_1)(x_3 - x_2)}y_3?$$

2. What degree is this polynomial? *(If the students decide that this represents a polynomial of degree 3, the instructor asks if this is always the case, and prompts them to notice that for some sets of data it may represent a polynomial of lower degree.)*

- **Task Example 4**

  Course: Linear Algebra

  Concept: Inverse matrices

  Task: 1. Decide with your group on a secret message. (It is a good idea to keep your message relatively short, say, one or two short sentences.) Associate each letter of your secret message with its position in the alphabet, and assign the number 27 to any blanks. Use the matrix:

  $$A = \begin{bmatrix} 1 & -2 & 2 \\ 3 & 1 & 1 \\ 2 & 0 & 1 \end{bmatrix}$$

  to encode your message. If the number of characters in your message is not a multiple of three, you will have to pad your message with one or two blanks at the beginning or the end.

  2. Exchange your encoded message with another group, and try to decode their message.

  (This kind of activity engages the students in conversations between the groups as well as within the groups. In talking with each other about the strategies that they are using to encode and decode their messages, the students are engaging in a "think aloud" process advocated by Davy (1983) (see Section 2.2.4).)

- **Task Example 5**

  Course: General Statistics

  Concept: Least squares regression line

  Task: 1. *(A scatterplot with 5 points is drawn on the blackboard, and
                a computer with a statistical package is available.)* Use your own
                rules to find a line of best fit. Compare your line with the lines
                determined by the statistical package.

              2. *(The least squares method is described if no group suggests it.)*
                Use the least squares method to determine a line of best fit for
                the scatterplot on the blackboard. Compare this line with the
                line determined by the statistical package.

- **Task Example 6**

  Course: Finite Mathematics, Mathematical Modeling

  Concept: Linear programming, graphical method

  Task: *Power Machines, Inc.* is a small manufacturing firm that produces
  motorcycles and snowmobiles. Two machines are used in the production of both of these vehicles. Producing a motorcycle requires two
  hours on machine A and two hours on machine B; producing a snowmobile requires one hour on machine A and four hours on machine B.
  Machine A can be operated up to ten hours per day, while machine
  B can be operated up to sixteen hours per day. Each motorcycle that
  is sold brings a profit of \$500, and each snowmobile brings a profit of
  \$700. Assuming that all the motorcycles and snowmobiles that can
  be produced will be sold, how many of each should be manufactured
  to maximize the company's profit?

  *(The students have already made a graph of the feasible region for this
  problem situation in the preclass activity.)*

  Consider the objective function $P(x, y) = 500x + 700y$. The plant
  manager is interested in finding the values of the decision variables x
  and y that will maximize the profit. Add the following lines to your
  graph of the feasible region for this problem situation:

  $500x + 700y = 0$
  $500x + 700y = 2400$
  $500x + 700y = 2600$
  $500x + 700y = 2900$
  $500x + 700y = 3100$
  $500x + 700y = 3400$

The following points are to be drawn out in the class discussion as the students work on this task:

- *The lines representing various values of the objective function are parallel, and one of these, $P(x, y) = 3400$, intersects the feasible region at one of the vertices. In fact, this vertex point, $(4, 2)$, represents the values of the decision variables that give the maximum profit.*

- *The maximum profit occurs at a point (or set of points) where two conditions are both met:*

  * *the objective function intersects the feasible region, and*
  * *the objective function has the largest possible x- or y-intercept.*

  Thus, if there is an optimal value (maximum or minimum) of an objective function, it must occur at one (or more) of the vertices of the feasible region.

- *Since this graphical method of solving linear programming problems has some limitations, there are other methods (e.g., the simplex method) for solving problems involving more than two decision variables.*

The interaction of each student with the problem and the social interaction among the students during the preclass activity and class task helps to create an environment in which each student has the opportunity to make the mental constructions necessary to assimilate this new mathematical experience. After doing this class activity, the students should be ready to give an informal proof of the theorem that the optimal values of the objective function occur at the vertices of the feasible region (see Sections 2.3 and 2.4).

- **Task Example 7**

  Course: Intermediate Algebra

  Concept: Quadratic models

  Task: 1. Find the x-coordinate of the vertex of the parabola represented by the quadratic function:

  $$y = ax^2 + bx + c.$$

  The students may have observed that the vertices of the several parabolas that they graphed in the computer laboratory (see page 37) all had the same x-coordinate, and that the x-intercepts of each of these parabolas were symmetrically located with respect to the vertex. Recalling and articulating these observations will help them realize that the constant c has no effect on the value of the x-coordinate of the vertex. By working with the simpler expression $y = ax^2 + bx$, the students can construct an expression for the x-coordinate of the vertex in terms of the coefficients a and b.

2. Farmer MacDonnell wants to use 800 feet of fencing to enclose a rectangular plot of land next to a river. (No fence is needed along the side adjacent to the river.) Determine the dimensions of the largest rectangular plot that can be enclosed. What is the maximum area that Farmer MacDonnell can enclose with this much fencing material?

- **Task Example 8**

Course: Linear Algebra

Concepts: Linear combinations, linear independence, spanning set, basis of a vector space

Task: A set of vectors, $S = \{v_1, \ldots, v_n\}$, is called a *basis* for $\mathcal{R}^n$ if both of the following conditions hold:

  – S is linearly independent in $\mathcal{R}^n$.
  – Every vector in $\mathcal{R}^n$ can be expressed as a linear combination of the vectors in S.

1. Show that the set:

$$\{[1,0,0,0], [0,1,0,0], [0,0,1,0], [0,0,0,1]\}$$

is a basis for $\mathcal{R}^4$. In fact, this is called the *standard basis* for $\mathcal{R}^4$.

2. Determine whether the set of vectors $\{v_1, v_2, v_3, v_4\}$ that you worked with in the activities (on page 37) is a basis for $\mathcal{R}^4$.

3. Prove that the following set of vectors is a basis for $\mathcal{R}^3$:

$$\{[2,1,-3], [1,0,4], [-1,2,1]\}.$$

(Concepts of spanning set, linear independence, basis, and dimension of a vector space can be introduced in class discussion. The students have already worked with a set of vectors in the activities (on page 37) that form a basis for the vector space $\mathcal{R}^4$. Since they have already done all the calculations necessary, they can focus on the new definitions; they have already worked with an example that illustrates the definition. The discussion following these class tasks can lead the students to an informal proof that every basis for a particular vector space must have the same number of vectors.)

4.3 Exercises after a concept is studied

After the students have the opportunity to construct mathematical ideas by attempting the preclass *activities* and have worked with these concepts by doing class *tasks*, they are given a set of *exercises*. These exercises are standard problems designed to reinforce the concepts being discussed and to challenge the students' thinking. In addition to regular drill-and-practice problems, the exercises might include both writing problems and computer-based problems.

Usually group members meet to develop their own plan to do homework assignments. This plan includes what to do, and who will do what. We encourage the students to do the homework assignments with their groups. Ideally, the students discuss each exercise as they work through the assignment. Sometimes the group will split into subgroups that each take responsibility for part of the assignment. This is less than ideal, and we remind students that each person needs to understand each of the homework exercises.

At those times when, as a matter of expediency, individuals work alone on certain exercises without getting back together to discuss the work with their group, they "short circuit" the learning process. We say more about this problem in Chapter 6 (see Section 6.1.2).

If an assignment involves computer-based problems, members of a group need to meet where the software is available. In addition to working during any scheduled computer laboratory sessions, groups are expected to meet at other times to complete any such problems. Thus their plans to complete an assignment are complicated by the constraint imposed by the open hours of computer laboratories (unless one or several of the group members have a computer with the appropriate software at home or in the residence hall).

No matter how a group does the exercises (in the whole group, in subgroups, or individually), the group is expected to meet to discuss all solutions and to prepare the solutions for submission. As the students edit each other's work, they continue to discuss the solutions and make conjectures. Many of the authors and respondents to the survey require a cover sheet to be submitted with each homework assignment; a student's signature on this cover sheet indicates that he or she understands each solution and has contributed to the work.

Some instructors believe that in addition to group homework, students should work on individual homework to ensure the active participation of each student. In this case, students have both group and individual homework assignments.

4.3.1 Examples of exercises

The exercise examples given below are related to the activities (see Section 4.1.1) and tasks (see Section 4.2.1) with corresponding numbers.

- **Exercise Example 1**

  Course: Precalculus, Trigonometry

  Concept: Identities involving sine and cosine functions

  Exercise: 1. Evaluate each of the following expressions:
  - (a) $\sin^2(0) + \cos^2(0)$
  - (b) $\cos(2x)$ given that $\cos(x) = 0.5$
  - (c) $\sin(x - \frac{\pi}{2})$ given that $\cos(x) = 0.25$

  2. Sketch a graph of the function g given by:

  $$g(\theta) = \sin^2(\theta) + \cos^2(\theta).$$

  What is the domain of this function?

  3. Show that the following expression is an identity:

  $$\cos(2x) = 2\cos^2(x) - 1.$$

- **Exercise Example 2**

  Course: Calculus I

  Concept: Properties of exponential functions and the number e

  Exercise: 1. Find the derivative of the function represented by each of
  the following expressions: e^{2x}, $x^2 e^x$.

  2. Write a formula for the derivative of the function g given by:

  $$g(x) = e^{f(x)}.$$

- **Exercise Example 3**

  Course: Mathematical Modeling, Numerical Analysis

  Concept: Interpolation and curve fitting

  Exercise: 1. Find a polynomial of the lowest degree possible to fit each
  of the following sets of points:
  - (a) $\{(1, -1), (-2, 2), (-5, 0)\}$
  - (b) $\{(4, 3), (1, -1), (-2, 2), (-5, 0)\}$
  - (c) $\{(1, -1), (3, -3), (-4, 4)\}$

  2. If you are given a set of n points, can you always fit a polynomial
  to these points? What is the degree of this polynomial? Is it
  unique? Explain.

- **Exercise Example 4**

  Course: Linear Algebra

  Concept: Inverse matrices

  Exercise: 1. Use the matrix:

  $$A = \begin{bmatrix} 1 & 2 & -2 \\ -2 & -3 & 5 \\ -1 & 2 & 7 \end{bmatrix}$$

  to encode the message: "Reinforcements are coming."

  2. Decode the message:

  1, 38, 179, −15, 44, 82, −11, 43, 115, 33, −61, −12, −3, 47, 194.

  3. Show that the following matrices are inverses of each other:

  $$\begin{bmatrix} \cos(x) & \sin(x) \\ -\sin(x) & \cos(x) \end{bmatrix} \quad \text{and} \quad \begin{bmatrix} \cos(x) & -\sin(x) \\ \sin(x) & \cos(x) \end{bmatrix}$$

- **Exercise Example 5**

  Course: General Statistics

  Concept: Least squares regression line

  Exercise: 1. Use the method of least squares to find the line of best fit *(for some simple given data set)*. Show all your calculations.

  2. Use a statistical package to find the line of best fit *(for the same data set)*.

- **Exercise Example 6**

  Course: Finite Mathematics, Mathematical Modeling

  Concept: Linear programming, graphical method

  Exercise: The *Furniture Mart* is getting ready for its annual sale on recliners and love seats. The storeroom has only enough space for 45 items. It takes one hour to unpack and set up a recliner, and two hours to unpack and set up a love seat. The store manager has 90 employee-hours available to unpack and set up these items. Each recliner sells for $350 and each love seat for $500. How many of each should be sold to maximize revenue?

- **Exercise Example 7**

  Course: Intermediate Algebra

  Concept: Quadratic models

  Exercise: 1. The cost of operating *Jimbo's Car Wash* is given by the
  function C defined by:

  $$C(x) = 0.5x^2 - 20x + 600,$$

  where x is the number of cars served daily. How many cars should
  be served daily to minimize the operational cost of this car wash
  business? What is this minimum cost?

  2. Suppose that the approximate number of people, N, in a department store can be modelled as a quadratic function of time, t.
  Let $t = 0$ correspond to 9 a.m. when the store opens, and assume
  that the number of people in the store at both opening and clos-
  ing (9 p.m.) times is 0. If the maximum number of people in
  the store on a particular day was 216, find the function $N(t)$ and
  determine the approximate number of people in the store at 11
  a.m.

- **Exercise Example 8**

  Course: Linear Algebra

  Concepts: Linear combinations, linear independence, spanning set, basis of a vector space

  Exercise: 1. Let $M_{m \times n}$ be the set of all $m \times n$ matrices, and P_n be
  the set of all polynomials of degree n or less. Find the standard
  basis for $M_{3 \times 2}$ and for P_3.

  2. Which of the following sets of vectors forms a basis for $\mathcal{R}^4$? In
  each case, explain why the set is or is not a basis.
     (a) $\{[1, 1, 2, 2], [1, 2, 0, 1], [0, 1, 1, 0]\}$
     (b) $\{[1, 1, 2, 2], [1, 2, 0, 1], [0, 1, 1, 0], [-1, -1, -2, 1]\}$
     (c) $\{[1, 1, 2, 2], [1, 2, 0, 1], [0, 1, 1, 0], [2, 4, 0, 2]\}$
     (d) $\{[1, 1, 2, 2], [1, 2, 0, 1], [0, 1, 1, 0], [-1, -1, -2, 1], [3, 2, -1, 4]\}$
  3. Show that the set $\{f, g, h\}$ where the functions are defined by:

  $$f(x) = 2x^2 + x, \quad g(x) = x^2 - 2, \quad h(x) = 3x^2 - 2x + 2,$$

  is a basis for P_2.

4.4   Group activities in the computer laboratory

In some colleges, certain mathematics courses have a laboratory component in addition to regular classes, while at other colleges it is impossible to schedule laboratory sessions. In the latter case, an instructor may conduct some of the classes in a computer laboratory or computerized classroom. We will describe the use of cooperative learning in both of these models.

4.4.1   Scheduled computer laboratory sessions

A course might meet each week for two to four hours in the classroom and one to three hours in the computer lab. The computer lab is supervised by the instructor, who might be assisted by a graduate or undergraduate student. The basic goal here is to integrate technology into a cooperative learning environment, and to create a situation in which the computer is used as an additional "group member" — providing another form of "conversational feedback" (see Section 2.2.4).

The students work on previously assigned computer-based problems in the computer lab, where members of each group sit together and share at least one computer. A group may use several computers, especially when doing lengthy calculations or when using more than one kind of software. Students may have access to both a symbolic computer system and a mathematical programming language; in addition, they may have a dynamic geometry system, spreadsheets, and/or various graphing tools.

Members of each group need to develop their own plan for working in the computer laboratory. To use the lab time productively each group is advised to read and think seriously about the assigned problems (even to try to devise a plan to solve each problem) before coming to the computer lab. Then the lab time can be used to carry out their plans using computers. In this environment group members often detect and correct errors in their plans to solve a problem. Members of a group investigate the computer output, analyze the graphs produced, and make inferences and conjectures. Groups are expected to either complete assigned problems during that session in the computer laboratory, or to meet outside the scheduled laboratory session to complete the assignment.

It is not unusual to observe the students interacting with the computer as though it is an additional group member. While working in groups in the lab, students appear to be less afraid of making mistakes or coming to wrong conclusions. The authors and the respondents to our survey have observed that students who may not speak out in large group discussions appear to feel more comfortable asking questions and explaining the material to each other in their small groups; some students seem to find it easier to talk about a result that the computer gave than about their own mathematical work (see Section 2.2.4).

The instructor should encourage active and full participation of all group members in all group activities. In the computer lab, the instructor is a facilitator who supports students with hints and ideas. It is not the task of the instructor to provide solutions at this time. The students are to find their own mathematical solutions, or to figure out why they do not obtain the outcomes they expected. The instructor may ask for the attention of all groups to discuss a common error among groups, to explain difficult notation, or to give a helpful hint.

4.4.2   Classes conducted in a computer laboratory

In some schools mathematics courses are not considered as laboratory courses; it is not possible to build computer laboratory times into the course schedule. However, if computers and cooperative learning are integrated into a mathematics course, the instructor may conduct some of the classes in a computer lab or in a classroom equipped with computers. In this case, the instructor has more control over the time allocated for groups to work on activities as well as the time designated for class tasks and discussion. Some activities and tasks may be computer-based problems, while others may be regular paper-and-pencil problems. The computer tasks are designed to promote conceptual understanding and to form a basis for class discussion. The instructor can use the computers as a powerful teaching tool in a cooperative learning environment.

The students work on these tasks in groups when group answers are required. In solving a computer-based task, each group member is expected to understand the problem, devise a plan to solve it, and carry out the plan. Frequently students carry out the devised plan together; sometimes two subgroups, acting independently, carry out the plan or part of it. After a group discussion, students formulate their group answer. During the group work on the tasks, the instructor moves from one group to another to help students in various ways as discussed above. The instructor also conducts any class discussion.

4.5   Group preparation for quizzes and examinations

In cooperative learning, there are three ways to deal with tests: all tests can be group tests, all tests can be individual tests, or there can be a combination of group tests and individual tests. The pros and cons of each situation are discussed in Chapter 5 which deals with evaluation and assessment. In that chapter we also discuss the reasons for our belief that it is more beneficial to have a combination of both group and individual tests.

At times it is helpful to raise the level of concern that students have about their own understanding of the material, while at the same time providing individuals with the "safety net" of group interaction. Group quizzes can be used to do this. Quizzes may be similar to class tasks except that group solutions are collected and graded. After a few moments of silence so that students can think about the problem individually, group members discuss their problem-solving strategies, compare their answers, and prepare a group response to submit.

Some of the authors and respondents to our survey have also given group tests. Several days before a group test, the instructor provides students with a list of the topics to be covered on the test and discusses with them the format of the test. A group test may differ from an individual test that covers the same topics by including some nonroutine, challenging problems for which the contribution of each member is needed and group creativity is required. The instructor should discuss with students various strategies that can be used during a group exam. The instructor can arrange for a review session with groups if time permits. It is assumed that the grading system has been discussed at the beginning of the course.

Members of each group are encouraged to devise their own plan to study and review the material for a group exam. If the group members cooperate in preparation for the exam, all group members benefit from their group review sessions. It is expected that students will make their best efforts on a test since the group grade affects their final individual grade.

Even if all the tests are individual tests, some of the authors and respondents to our survey have noticed that some groups get together before each test to study and review the material for the test. Some students seem to have found that this is an effective way to do at least a part of their preparation for tests and examinations.

All of this is discussed in greater depth in Chapter 5, and potential problems with group tests are discussed in Chapter 6.

4.6   The time students spend on group activities

The authors are often asked whether students must spend much more time on a course with cooperative learning than on a course where the students work individually. It is difficult to answer these queries from our own experience because we introduced cooperative learning at the same time that we made many other significant changes from our traditional lecture courses. Is it the cooperative learning, or is it the use of computers, the laboratory assignments, or the graded homework that has affected the students' time?

We might ask what overhead is imposed on students by cooperative learning. This question is analogous to the question asked about time devoted to learning a symbolic computer system or a programming language in a mathematics class. In a first course with a cooperative learning component, there may be a discussion or reading assignment to introduce the students to the expectations and benefits of cooperative learning. Later on there may be occasional group or individual conferences with the instructor to discuss group interaction. Throughout the course the instructor may occasionally spend a few minutes of class time reinforcing productive group behaviors or discussing any general problems observed. At the end of some laboratory sessions or classes, students may take a few minutes to arrange a group meeting. Students need to organize some of their study times to meet with others in their group; there may be some travel time involved in getting to these group meetings. Some students take time to think about the group interaction and to write about their group experiences in a journal. All in all, the overhead directly related to cooperative learning is relatively small.

On the other hand, the students do spend significant amounts of time on the activities and exercises. When working in their groups, the students do not run out of ideas and give up on a problem as readily as when the individuals work in isolation. When the group assignments are collected and graded (see Section 5.3), the students feel that the instructor is recognizing and rewarding their efforts. Thus the social interaction and support from their group members as well as the recognition of the group's efforts implied by grading group assignments encourage the students to schedule and spend a lot of time engaged in mathematical pursuits. Perhaps some of the benefit of cooperative learning is attributable to the additional productive time students spend studying outside of class.

4.7 Chapter Summary

In this chapter, we have discussed various group activities that may be used in a cooperative learning model. The groups work together and interact in supervised computer labs, in classes, and outside of classes and labs. They take group exams for which the group may submit only one copy of their solutions.

Before studying new concepts, the students work in groups on carefully designed *activities* that allow them to explore the concepts and construct mental images for them. These preliminary activities may be computer-based problems that the students work on in a scheduled, supervised computer laboratory. Each group develops its own plan of action in working on *activities*, class *tasks*, and homework *exercises*. We discussed possible group plans.

Group members sit together in classes. The instructor introduces new concepts in an intuitive way and asks students to work in their groups on class tasks. The *tasks* are designed to promote conceptual understanding and to help students develop the necessary mental constructs. In class discussion, the students reflect on their own group and other groups' experiences from preclass activities and class tasks. The similarities and differences of group answers form a basis for formal discussion of the new concepts.

After the new concepts have been introduced in preclass activities and have been discussed in class tasks, the students are ready to work on assigned *exercises* to reinforce the concepts being discussed and to test their understanding. In addition to regular drill-and-practice problems, the exercises may require some written explanations as well as the use of computers.

In any cooperative learning model, the students expect some group exams. We discussed common group action in preparing for and taking group exams. (We say more about group examinations in the next chapter.)

In general, in a cooperative learning model, traditional lecturing is minimized and the students become active participants rather than passive listeners. The instructor introduces the new concepts intuitively, monitors group work in classes and computer labs, initiates class discussions, and formulates or ties together ideas.

Chapter 5

How Do We Test and Assess Our Students?

There is considerable evidence that, whatever is intended by the designers of a curriculum and whatever is expected by the instructor who implements it, what actually happens in a course is largely determined by the students' perceptions of how they are going to be assessed. Examinations and grades are still a very large part of the educational experience at all levels, and it is not unreasonable for students to act in ways they think will lead to the highest grade possible in a course.

It follows then that assessment can be used as a way of influencing students to work cooperatively. If students are expected to work cooperatively in groups as they learn the material, then it is reasonable for the assessment of their learning to have a meaningful relation to their group work. More directly, if, for example, examinations can be set up so that students can expect to do better if they work cooperatively, or even if their group work is assessed directly, then it can be expected that the cooperative learning aspect of a course will be taken very seriously by the students. In other words, in addition to using assessment to evaluate individual students, we think that it is important to use assessment as a motivating force for collaborative work.

This raises another issue about which only a little can be said because we know only a little. In our observations of students working in groups, the authors have come to wonder if there is not a notion of *group knowledge*. Is it possible for a group to understand a piece of mathematics which no member of the group appears to understand? What could this mean? We are not thinking only of the phenomenon (which occurs often) that various members of a group understand various parts of a complex idea. We are wondering if the entire set

55

of activities one might perform to assess understanding of a particular concept could be applied to a group with a result that is much more positive than if it were applied to any individual member of the group. As we said, not very much is known about this possible phenomenon. We do know a little about the way in which a group might come to understand a concept as opposed to how an individual might do so (Vidakovic, 1993).

5.1   Testing in general

We want to make some general comments on testing before discussing testing in a cooperative learning situation. The normal practice in writing an examination is to determine the set of mathematics to be covered and then to design questions that try to determine what portion of that material a particular student has learned. This portion is expressed as a percentage and, eventually, as a grade. When we curve the scores on an examination, then we are comparing, for that material, what a student knows relative to what other students know. We are at a total loss in trying to understand how the comparison of one student's knowledge with that of the other students' knowledge has any real importance in terms of learning. Not only does it set up an unhealthy situation in which some students' success depends to some extent on some other students' failure, but it forces a conflict of interest between the instructor and the students. The students would like to get A's in the course. Almost all of them want this. When grades are curved, the instructor guarantees *a priori* that this cannot happen. It seems very difficult to establish a community of interest among students and instructors in such a competitive atmosphere.

We also question the practice of measuring a particular student's knowledge against the totality of the mathematics contained in the material covered by an examination. Taking the long view, we recognize that the logical dependence of one mathematics topic on others that, presumably, appear earlier in the curriculum, is not really reflected in how learning takes place. Students simply don't learn the prerequisites first and then the logical consequences. It is much less linear and, at any given point, students are struggling to understand both the present topic and the earlier ones on which it depends. Given this, it is reasonable that the study of mathematics involves a continual returning to the same topics over and over.

So it does not really matter so much if a student misses a point. He or she will have a chance to pick it up later. What really matters is how much the student has learned, how far he or she has come from a previous state of understanding. If, every time we assess a student, we find that he or she has learned a great deal that was not understood at the time of the last assessment then, in the long run, this student is going to do very well in mathematics (Page, 1979a).

We implement this last point of view by engaging in the following thinking about examinations. We look at the set of mathematics covered by the exam and we ask what are the most important, deepest, and difficult ideas, techniques or applications in mathematics that we think the students have learned. Then we ask the most difficult questions about these points that we think the students can handle. If there is a point that we are fairly certain has been missed by most students, we omit it from the current assessment and give the students further opportunities to learn any such crucial point. In this way, we try to find out what the students have learned, not what they have failed to learn, and we hope to encourage rather than to frustrate the students. Of course, part of the evaluation of the effectiveness of the course is based on how much material can be covered in exams designed this way and how hard the problems are — assuming that we were right and that the students performed well on the examination.

5.2 Testing and cooperative learning

Since we believe that cooperative learning promotes individual learning and students believe that the examinations indicate the important emphasis of a course, we advocate using the testing situation to foster cooperation. The use of group examinations, group preparation for tests, and group averages on individual tests stimulate the team spirit of the students in a cooperative learning group. The students are encouraged to take responsibility for the learning of other members of their group, and every student benefits from the ensuing interaction.

5.2.1 Group examinations

By a group examination we mean an examination where the members of each group work together on the questions and each group submits one solution for each problem. Furthermore, every student in a group is given the same grade. We recommend that a group examination be a somewhat longer and more challenging test than would be given as an individual test. In fact, some of the respondents to our survey have totally eliminated time limits. Exams are given in the evening and, within reason, students are allowed to work on the questions for as long as they wish. Of course, this requires a little extra work on the part of the instructor to make sure that a room is available and that students have enough advance notice to be able to arrange their schedules to accommodate evening or weekend exams. Some of our respondents reported that such scheduling could be a major difficulty for students and, in some cases, it was not possible to have exams outside of class time.

As mentioned earlier, we think that the questions should be designed to find out what the students know rather than to expose gaps in their knowledge so that the students' achievements are recognized and rewarded. Especially on a first group test, we believe that it is essential that serious groups have a chance at real success after an intense effort.

We suggest that instructors carefully prepare students for a first group examination by discussing test-taking strategies. Some special strategies for group tests include giving individuals time to think and get a problem fully in mind before discussion, comparing and checking answers when one or two students have finished a problem, and working in pairs (in a group of four) on different questions to make sure that all questions are attempted. Actually, the students probably have discovered and used the same strategies to effectively complete their group assignments. In an ideal situation, all members of a group understand and agree upon the group's solution to each problem.

Group testing in a mathematics course is a new experience for most students — and for most instructors. In the educational setting, we are accustomed to testing individual students on their mathematical knowledge. Yet, when these same students enter the working world, they are expected to use their individual knowledge in conjunction with the expertise of others working on a project. They must be able to integrate the knowledge of the whole group in order to solve a problem, and in doing so, individuals acquire much of the knowledge previously held by others in the group. Thus, although group testing is different from the familiar testing situation, it more accurately reflects the way in which students will be expected to function once they leave academia.

We can consider the analogous situation of an orchestra. The musicians in an orchestra must practice on their instruments individually, and they must learn their parts as well as they can while practicing alone. But when they come together for rehearsals, the individuals learn even more from each other as they play through the pieces together. Stand partners and members of a section freely exchange ideas on the technical challenges and interpretation of the music. There is growth and refinement of individual knowledge since musicians certainly learn quickly through their own observations where they count incorrectly or play out of tune. And an individual picks up any missed nuances from the conductor, those sitting nearby in the same section, and other sections of the orchestra playing similar themes. A tremendous *esprit de corps* develops as a concert approaches, and all hope that the group's presentation is well received by the audience and meets their own standards of performance. Each individual works to make the playing not only correct but elegant and musical. The calibre of performance depends on the individuals' contributions as well as on the way each musician interacts with the whole group. The final performance of a challenging program is certainly a group examination, and the preparation for that examination advances the individuals' knowledge and understanding.

There is a similar analogy with team sports, where the individuals as well as the interaction of the team members are tested in a game against another team. Thus, although our students find group testing in mathematics courses a foreign notion, many of these same students have experienced a form of group testing by participating in musical organizations and sports teams. The main difference in these experiences is that the group's performance in the year's concerts or competitions is evaluated. Although there are rewards and honors for outstanding individual accomplishments (and consequences for poor individual performance), the primary evaluation is that of the group rather than of the members of the group. Each individual's performance is not necessarily formally assessed as is required at the end of the semester in a mathematics course.

In the past few years, about seventy-five per cent of the respondents to our Survey on Cooperative Learning included at least some group tests in their courses, but very few have tried a group final examination. Group tests reported by our respondents include the major exams that occur two or three times during a semester as well as short quizzes given during the regular meetings of the class. All the instructors support the concept of individual accountability by including at least some individual tests in the course, and the most common plan is one group test out of three tests. Some others give two group tests and two individual tests. One instructor gives four tests, each with an individual part and a group part. The authors recognize that there are many different possibilities for effectively testing with a mixture of group and individual tests. We think that students' learning can be enhanced by including both kinds of tests in a course using cooperative learning groups.

5.2.2 Questions on group tests

One important feature of group tests is that it is possible to include more difficult questions and questions that require students to delve more deeply into a mathematical situation than is possible in ordinary tests. This is especially true if, as was suggested above, time is not a major factor for students taking the test.

Some of our respondents have reported on the effect of making the very first major test in the course a group test with difficult questions. Their experience is that the students are fairly successful, on average in the 70 - 80% range. Students concentrate on their success in solving problems more difficult than those they usually see on a mathematics test. They are less concerned with the possibility that they had done well only because they had help from their group members or a lot of time to solve the problems. They are proud of their achievement, and often this experience is a turning point in the course as students move from skepticism to beginning to develop ownership in a totally new, but apparently effective, method of learning mathematics.

To make clear what we mean by questions more difficult than on standard examinations, we provide some representative examples. The reader is reminded that these are not the exceptional questions on the test to separate the A's from the B's but that all of the questions are on this level.

Quiz in Calculus I

The following question was given as a 15-minute quiz about midway through the first semester of calculus. (Note that in this particular calculus program, about half of the first semester is restricted to precalculus with topics such as number systems, sets, functions and limits.) Although the question has the form of a multiple choice question, the requirement that the student explain her or his choice makes it a much more serious question.

Directions: *Take about five minutes to think about this question on your own. Then get together with your group to talk about this problem. Your group is to turn in one response to these questions.*

The *distance* traveled by a falling object is approximately $9.8t^2$ meters t seconds after it is dropped. As it falls, an object will gather speed; that is, it falls faster and faster until it meets some kind of obstruction (for instance, the ground). Which of the following expressions would give the *average speed* of the falling object over the interval from time t to time $t + 0.01$ seconds after it is dropped? Explain your choice.

1.
$$\frac{9.8t^2}{t + 0.01}$$

2.
$$\frac{9.8(t + 0.01)^2}{t + 0.01}$$

3.
$$\frac{9.8(t + 0.01)^2 - 9.8t^2}{0.01}$$

4.
$$\frac{9.8(t + 0.01)^2 - 9.8t^2}{t}$$

Group test in Calculus I

The following questions are from a group examination that took place about six weeks into a 15-week first course in calculus for engineering, science, and mathematics students. Again, this is still within the period during which the topics are mainly precalculus.

1. Given the following definitions of the functions f and g, define the composition $f \circ g$.

$$f(x) = \begin{cases} x^2 & \text{if } x \geq 0 \\ |x|(5-x) & \text{if } x < 0 \end{cases}$$

$$g(x) = \begin{cases} \frac{1}{5-x} & \text{if } x \neq 5 \\ 3 & \text{if } x = 5 \end{cases}$$

2. Suppose that the function f is defined by:

$$f(x) = \begin{cases} x^2 & \text{if } x \leq 3 \\ -x^2 & \text{if } x > 3 \end{cases}$$

and you are trying to determine the limit of this function as x approaches 3.000001. Which of the two branches of the function will you be more interested in and why?

Group test in Calculus II

The following questions are from a group examination that took place about six weeks into the 15-week second course in calculus for engineering, science, and mathematics students. At this point, the study of differentiation has been completed and the students are well into their study of integration.

1. Use the definition of the natural logarithm as a definite integral with variable upper limit to show that:

$$\ln(ab) = \ln(a) + \ln(b).$$

2. Given that the function sinh (hyperbolic sine) is given by:

$$\sinh(x) = \frac{e^x - e^{-x}}{2},$$

derive a formula for the derivative of the inverse of this function.

Group test in Calculus III

The following questions are from a group examination that took place about seven weeks into a 15-week third course in calculus for engineering, science, and mathematics students. The emphasis in this course is on functions of several variables.

1. Consider the function represented by the pair of parametric equations

$$x = a(t), \quad y = b(t).$$

   Find an expression which represents this function which has the form $y = c(x)$, where c is a function of one variable, and derive formulas for $c'(x)$ and $c''(x)$.

2. Let F be a function of three variables, so that the expression $F(x + y, x - y, y)$ represents a function of the two variables x and y.

   Express this situation in terms of partial functions in one variable, and find formulas for the first two partial derivatives of the function of two variables.

3. Which of the following functions is continuous at $(0, 0)$?

$$f(x, y) = \begin{cases} x & \text{if } y = x^2 \\ 0 & \text{if } y \neq x^2 \end{cases}$$

$$f(x, y) = \begin{cases} 1 + xy & \text{if } y = e^x \\ 0 & \text{if } y \neq e^x \end{cases}$$

Explain your answer.

These questions are representative of questions students can answer on group tests. The examples illustrate that many of our questions require the application of conceptual understanding to particular situations.

5.2.3   Group preparation for examinations

The authors find that many mathematics students, especially freshmen, need guidance in preparing for their examinations. Whether an examination is to be taken by individuals or by groups, the students benefit from guided preparation within the group structure. Although many students understand much of each day's work, most students need to carefully review the work of a few weeks to understand the logical connections and to know all the concepts well enough to apply them on the day of an examination. Most students want to succeed, but many of them don't know how to study, and, in particular, they don't know that they should write out solutions to problems as part of their studying. In our experience, by the time of the first examination, the students in most groups have bonded together closely enough that they are willing to help each other prepare for the examination. And, in fact, many students have come to rely on their groups for encouragement and emotional support as well as mathematical help.

Some instructors hand out a detailed list of the topics to be covered by the examination several days beforehand. Then the suggestion is made that the groups meet outside class to study each topic and the related homework and class work. The students who study together have an opportunity to ask questions in a comfortable setting and to refine their thinking as they explain their ideas to each other.

A group problem session during the class period before an examination has been successfully tried in some courses. Students are encouraged to study for the test before the review and to plan their other work so that they can devote as much time as possible to studying mathematics just after the review. On the day of the review, they sit in their groups as they work through problems together. They are all given questions similar to those that will be on the exam, printed in the format that will be used for the examination. These questions cover all the concepts in the material to be tested (but certainly they are *not* the examination questions). One member of each group assumes the role of tutor or coach. Each tutor is given one copy of the solutions, which are to be consulted only when the group is completely stuck or when the group has found a solution to the problem. On these days, students are highly focused on their mathematical work, and they generally interact with other members of their group in a very productive way. Since there are many more questions than can be answered during an hour, the groups continue to work on these questions after class. Many groups study much longer and more completely for the examination than they would without the review problems.

In any case, whether there is a structured review for an examination or not, the authors think that the students should be encouraged to arrange group study sessions in preparation for each examination.

5.2.4 Group grades and individual tests

When an individual examination is given, the authors still want all students to feel a part of their cooperative learning groups, and we especially want them to interact with their whole group as they prepare for an examination. One way to accomplish this is to include the evaluation of the group's performance as part of the grade of each student.

Some of our respondents use a method that has generated a certain amount of controversy. The idea is that among the major examinations in the course, in addition to one being a group examination, there is one for which the student receives two grades: her or his individual score and the average of the scores of the members of the student's group. Some instructors have philosophical objections to the grade of a student being so clearly affected by the work of others, and this method does put a certain amount of stress on the students — both those concerned about being pulled down and those feeling guilty about pulling the team down. But when it is successful, this method can make tremendous contributions towards developing in a group the feeling that each member of the group is responsible for the learning of all members of the group. This kind of team spirit can significantly enhance the students' learning experiences.

Other methods include rewarding the members of certain groups by adding a few extra points to each individual score whenever the group earns the highest average or an outstanding average, and, for examinations after the first, when the group attains the most-improved average. We suggest that the grading system be designed to motivate the students to study cooperatively for the individual examinations as well as the group examinations.

5.3   Other assessment of student learning

Whenever students do mathematics — in classroom discussion, computer laboratory assignments, and homework exercises as well as in examinations — they show that they know some mathematics. And, in fact, in these various situations, students exhibit different aspects of their knowledge. For example, in a course where the examinations are not taken in the computer laboratory, the examinations do not indicate whether a student can use a computer effectively to solve a mathematics problem. Further, closed-book timed written examinations do not show whether a student can read and understand written mathematics, solve a problem that takes a long time, or communicate and respond in discussions. We think that it makes sense to assess all the mathematics that the students do in a course, and to include this assessment as part of the total assessment in the course. We observe that students tend not to feel that any given assessment activity is a "do or die" situation when there are many scores of different kinds.

On the other hand, we know that most instructors cannot take time to do all this assessment for classes of any size even though they have far fewer papers to grade when each group of four makes one submission. We find that more advanced students, either undergraduates or graduate students, can handle the grading of written work, with the possible exception of examinations. We supply complete directions and solutions to undergraduate graders, but the graduate students usually write their own solutions cooperatively.

In Chapter 4, we described a classroom where the students discuss short tasks in their groups and then present their conclusions to the class. We use a simple grading scheme for such classes: the group is graded 0 or 1 for each response. The grade of 1 indicates a serious attempt to think through the problem, and a grade of 0 indicates that the group has absolutely nothing to say. On very rare occasions, for an outstanding response, a grade of 2 or 3 is awarded and joyfully told to the class. Once a response to the task is given by a group, other groups are given an opportunity to comment, and these groups are graded similarly. The instructor may announce the grade after each response, and either the instructor or a student assistant records the grade. Any student absent from class when her or his group is graded receives a grade of 0. We find that this approach is very helpful in stimulating discussion and improving class attendance.

Grades on computer laboratory and other homework assignments also influence the groups to seriously prepare their work in a timely fashion. As in the grading of classroom tasks, some instructors have developed a grading plan in which each problem is worth at most a few points, and many problems are worth only 1 point. The total grade for a problem is earned by a serious attempt that has some merit. This system avoids the necessity of making refined judgments on partial credit, which may be very difficult and time-consuming for a student grader. It also rewards groups for efforts that show the members are making progress towards understanding a concept. When corrections and comments are not written on the papers, we make solutions available, usually during laboratory periods. We believe that students are especially ready to learn after they have worked on a problem, received a grade that showed an incomplete or incorrect solution, and then are given the opportunity to compare their work with a correct solution.

In courses where students are required to keep journals, the instructor reads the new entries about once a week. Since journal writing is informal and amounts to a private communication between the instructor and each student, there is no attempt to correct the writing or the ideas. In a journal the student is learning to assess her or his own mathematical growth as well as experiences in the cooperative learning group, classroom, and laboratory. The instructor might give journal entries for the week a check if they show some thought. The instructor responds to any questions written in the journal, and he or she may also write comments and pose questions for the student to ponder. Some of us have found that students who have written inadequate entries one week respond well to being prompted by questions from the instructor. Good entries receive appropriate comments and praise. The number of acceptable journal submissions during the semester is used in the computation of the students' semester grades (see Appendix F).

5.4   Computation of the course grades

Once a decision has been made on what to grade, the instructor must determine how these grades are going to be used to compute the course grade for each student. We think that the students in a course involving cooperative learning, as in any course, should know right from the beginning just how the course grade will be computed. In other words, the students should know the relative weights given to any graded work such as examinations, the final examination, the classroom responses, laboratory assignments, and homework exercises. Then the students know the importance of the group work for their success in the course. Appendix F contains the grading schemes for several different courses.

In addition, the authors recommend making a conscious decision on the percentage of the total grade to be earned in group work. This percentage varied from 5% to 90% among the respondents to our Survey on Cooperative Learning. But most respondents were in the 20 – 50% range, and we think that it is reasonable to stay within this interval. Then the group work has a significant effect on the grade, and also the individual student is held accountable for learning the mathematics because, in our courses, much of the individual grade is earned in examinations.

5.5   Chapter Summary

The authors think that assessment can be designed to motivate students' behavior as well as to evaluate their work. We suggest that tests should give the students an opportunity to show what mathematics they know rather than to expose what they don't yet understand. We believe that the importance of cooperative learning can be communicated to students by including some group tests, group preparation for all tests, and group averages on individual tests. In particular, we advocate that the first examination of the semester be a group test, where the members of each cooperative learning group work together on the test questions and submit one solution to each problem. Questions on a group test can be more difficult than those on individual tests.

The authors also recommend that all of the students' work be evaluated and counted in computing the grades in the course. The students' contributions in the classroom, their work in the computer laboratory, their journal entries, their homework, and any other mathematical activity that is part of the course should be assessed by either the instructor or a student assistant. We think that this broad assessment shows the students the value of serious work in every aspect of the course. The authors and most of the respondents to the Survey on Cooperative Learning agree that the group work should count 20 – 50% in computing the course grade. Thus we believe also that a significant part of the students' grades should reflect the level of individual understanding and mastery of the mathematics covered in the course.

Chapter 6

How Might a Group Function?

Well. Or poorly. Or somewhere in between.

We have said a lot in previous chapters about why students should work in cooperative teams, how beneficial it can be to learning. We have also discussed the things that a group might do in a course organized around cooperative learning. The seven of us became involved in writing this book because we are certainly convinced that working cooperatively is a better way of doing most things in life — including learning mathematics. In other words, it is a great idea, a terrific theory about how a mathematics class can be organized.

Like all theories, all big ideas, cooperative learning is only as good as its implementation. Although cooperative learning can be a powerful pedagogical tool, it is not a panacea. Organizing the students into teams and structuring the course so that students do most of their work in their groups may not, in itself, lead to much improvement in learning. We will see in this chapter that there are many ways that a group and the individuals in it can function. Some of them seem to help the students learn and we call these *productive* ways. Other ways in which a group can function are not very helpful and might even make it less likely that students will learn. We call these *non-productive* ways. It would be naive to think that reality is anything other than various groups functioning in a mixture of productive and non-productive ways. When the latter are so predominant that it seems that very little learning is likely to take place, we say that a group is *dysfunctional*.

Up to this point, we have really said very little about the practical issue of getting a group to function productively. It is a good idea for students to work cooperatively so you should set up groups. Depending on how the group functions, learning might be enhanced or it might not. What we think is really important is that the overall environment of a course, including the structure in which it is organized and what is actually done in the implementation, provides abundant opportunity for influencing how the groups in a given class actually function. The instructor can have an effect, reducing non-productive and increasing productive ways of functioning. It is this reality that makes cooperative learning such a powerful pedagogical tool.

So we will explore in this chapter the various modes in which a group can operate and the difficulties or non-productive ways of functioning that might arise. Then we will talk about how the instructor might organize the course so as to reduce the occurrence of non-productive functioning and specific difficulties. More important however, since these troubles will occur at least to some extent no matter how the course is organized, we will talk about things that might be done to overcome them.

6.1   Groups can operate in various modes

6.1.1   How can you recognize a group's mode of operation?

The instructor must play an active role in becoming aware of how the groups are operating. It can be expected that, left to their own devices, students may let their groups fall into non-productive modes of operation. This is not entirely their fault. Our educational system is based on success in performance on examinations and there is an atmosphere in which it is acceptable to do the minimum amount of work, to take the easiest road, as long as you do well in the exams. Moreover, because our society pays relatively little attention to working cooperatively, as opposed to the myth of "rugged individualism", there is a natural tendency for student groups to evolve towards more individual and less cooperative modes of operation.

For an instructor to influence a group's mode of operation, it can be helpful to be aware of what that mode of operation is for a particular group. Almost all of the respondents to our Survey on Cooperative Learning reported using informal observations to see how groups were operating. If a course has a computer laboratory component then the instructor can watch the groups as they function. If the class has students working cooperatively on tasks, then the instructor can walk around the room and observe the groups. Casual meetings with students can provide an opportunity to ask about how things are going. Sometimes students' off-hand comments can give an indication of how things are going with the groups.

More formal ways of observing group operations can also be used. A favorite technique with our respondents was to schedule periodic meetings with the groups. Held during regular office hours, or during sessions held in the computer lab, these meetings may not use up much more time than is spent by an instructor normally in consultation with students. A skilled instructor can find out a great deal about how a group is operating and how the members of that group feel about it during a 20 – 30 minute session. Thus, in a class with ten groups, the instructor can easily meet with each group every couple of weeks.

Many of our respondents had students keep journals and turn them in periodically. Reading the journals is another source of information about modes of operation of a group. Some instructors occasionally distribute brief questionnaires that ask students for various kinds of information about their work in the course and, in particular, about how their group is functioning. An instructor can even tell something about group operations from the assignments that the students submit.

Thus, there are several ways in which an instructor can become familiar with the modes of operation of the groups in her or his class. Not only is this information helpful for the instructor in making adjustments, but students appreciate and respond positively to a strong show of interest.

6.1.2 What are the different modes of operation?

One of the first things observed by anyone who tries to use cooperative learning in their teaching is that organizing the class in groups does not automatically determine how each group will function. Our experience has been that when a course involves regular assignments to groups that stay the same, then each group will evolve fairly quickly into one or another of the following modes of operation.

- The entire group does every problem.

- The group divides up into subgroups coming together before, during and after attempting the assignment to compare results.

- Each individual does nearly every problem, and the group gets together to compare results before, during and after attempting the assignment.

Of course, it is rarely the case that a group will operate in a single mode at all times. There can be variation with time, type of activity (homework, exam, in-class tasks, etc.) or type of problem. In general, a group will use a mixture of the three modes, and, by observing the group, an instructor can get a feeling for the distribution that occurs.

6.2   Groups can run into many difficulties

We cannot repeat often enough that cooperative learning will not help very much unless it works in certain ways. There are many difficulties that can go so far as to completely eliminate any contribution to learning from this pedagogical device. Students will not learn much more than in standard courses if these difficulties are not avoided or overcome, at least to a reasonable extent.

The first step in avoiding or overcoming difficulties is to know what they are. We will describe the difficulties that we and our respondents have observed. You, the reader, may come up with additional problems in your own observations. (We hope that you will not run into too many that are not on our list.) In any case, being aware of what others have seen will help you as you work with your own students, if for nothing else than to let you know that you are not alone in having to deal with a particular problem.

We organize our list in several categories: difficulties connected with modes of operation, difficulties arising from organizational issues, difficulties connected with group dynamics, difficulties having to do with individuals, and difficulties having to do with the rest of the world.

6.2.1   Difficulties connected with modes of operation

We have listed the different modes above in what we consider, roughly speaking, to be in decreasing order of effectiveness, at least in terms of cooperative learning. Although we think that having the entire group do every problem is the mode in which the members of the group will have the richest learning experience, it is also the mode that takes the most time. Similarly, although dividing the problems among individuals almost totally negates the effect of cooperative learning, it will get the most work done in the fastest time — assuming the individuals are able to solve the problems.

Thus, the first difficulty connected with modes of operation is that groups will tend to function more in the modes lower down in our list rather than higher up. The key to effective cooperative learning is the coming together of individuals or subgroups to discuss and possibly revise the work after it has first been done. Herein lies the second difficulty in connection with modes of operation.

Using any other mode except the first, it is essential that groups do this coming together. This is a time when there is an opportunity to correct errors. It is also important to make sure that each member of the group has at least some understanding of the group work in which he or she was not personally involved. Group members must develop the habit of insisting that everything be explained, be justified, and that solutions are not accepted until everyone has some understanding of them. Finally, the coming together can lead to alternative approaches to problems being offered and discussed. "Why didn't you do it this way?" Right or wrong, the process of responding to this question is almost sure to be a rich learning experience for all members of the group.

Working individually or in subgroups and coming back together is a powerful way of learning that is on the level of full group work (where everyone participates in solving the problem). We don't know enough about how all this works and indeed, the separating and coming back might be even *more* effective. We refer to this as a difficulty (which was reported by about a third of our respondents) however, because in practice, once a group moves away from exclusive use of the "full group" mode, there is a tendency (justified by the lack of time that seems to pervade our culture) to skip the step of coming back together for a discussion with the full group. Then much of the value of cooperative learning is lost. Mature groups understand this and tend to employ a mixture of all three modes of operation, being careful to schedule time for "post-mortems." About a third of our respondents said that they encountered this particular difficulty.

6.2.2   Difficulties arising from organizational issues

College students need to learn a lot about organizing their lives and living up to their organizational commitments. Almost half of our respondents reported that students could not find a common time to meet; more than half reported as a difficulty that quite often students did not show up at planned group meetings; and sometimes, there was a reluctance to the very idea of the group meeting at times other than class or scheduled computer laboratory sessions, and in places other than on campus.

Very often we found that students would schedule the coming-together time just before the time at which the assignment was due. Enough time was usually allocated, but on many occasions a student was late; thus her or his contribution was unavailable for discussion, and the tardy group member also missed out on the rest of the discussion.

Another organizational difficulty that was occasionally mentioned was that a member of the group was not a native speaker of English, and the group had difficulty in communication. On the other hand, sometimes students report that the extra effort that is made to communicate in such a situation seems to challenge them to a deeper understanding of the material.

Finally, we should mention that the heterogeneity of groups is not always an unmitigated blessing. On occasion, the gap between the strongest and weakest members of the group is extremely large, and either one member loses patience or one member became discouraged — or both. Another form of this difficulty occurs in cases in which there are two group members very much stronger or very much weaker than the other members of the group. It is not easy in such situations to avoid the two forming a permanent subgroup that does not communicate very much with the rest of the group.

6.2.3   Difficulties connected with group dynamics

It is not surprising in a culture like ours in which children have very little experience (outside of competitive sports completely dominated by a "coach") with group activities, that the use of cooperative learning in college leads to a number of difficulties in group dynamics as several individuals attempt to function as a single unit.

Probably the most common difficulty that was reported by the respondents to the Survey on Cooperative Learning involved a single student dominating discussions. In the opposite direction there are often students with very little self confidence who do not speak out. It may not actually be important for every member of the group to have an equal amount of "air time." However, it is important for a group to function so that each individual has an opportunity to make her or his contribution to the fullest extent. For some this means speaking a lot and for others it means listening a lot. But the speaker needs to be sensitive to the needs of others; and the listener needs to ask questions, and even contribute thoughts on occasion. Operating so as to balance all of these conflicting needs is not easy for college students — or for anyone else!

Sometimes a group has one or another member who refuses to do her or his share of the work. For obvious reasons the group members are reluctant to take drastic steps such as not allowing a group member to sign (and get credit for) an assignment. The group has to learn that it is only through its own dynamics that such difficulties can be overcome. It must realize and act upon the dual notions that each member of the group benefits from the triumphs of each member and suffers from the failures of each member. Students will have both philosophical and practical objections to such ideas. However, in "buying into" a cooperative learning pedagogy, students must learn how to resolve this kind of conflict.

Less common but still important group dynamic difficulties can occur in groups with mixed gender or mixed ethnic origins. It is not unlikely that the conflicts of our overall society will be brought into a cooperative learning group and get in the way of students learning how to work together.

6.2.4   Difficulties with individuals

As with any complex system, a number of difficulties with individuals can occur. Most of our respondents mentioned the case of a student not doing her or his share of the work or failing to contribute to discussions in the group and in class; sometimes, an extremely good student feels held back by the group or a weak student has difficulty admitting that he or she is in trouble and asking the group to slow down; sometimes, a group member with the best of intentions to help instead stifles others; there are examples in which one member of a group is chronically absent, sometimes but not always for a good reason such as a serious illness; an individual might feel anxiety about computers or about mathematics; and finally there are any number of personality conflicts that arise.

All of these individual difficulties contribute to a mosaic in which cooperative learning, although potentially extremely effective in learning, might not achieve its potential.

6.2.5   Difficulties with the rest of the world

One final type of difficulty that is important is the negative feedback students may receive from their colleagues and even from other faculty members. Cooperative learning is considered by many to be a very promising pedagogical tool and to have the potential for making a big difference in the students' learning mathematics if we can learn how to use it effectively and avoid difficulties. Others disagree and express philosophical objections to this method directly to students; students awareness of others' disagreement can make it difficult to implement cooperative learning in the most positive way.

6.3   Dealing with difficulties

An important principle seems to be emerging from the experiences of the people who responded to our survey. Cooperative learning can be an extremely effective pedagogical strategy in helping students learn mathematics. There are, however a very large number of difficulties that could negate its effect. We don't know of very many experiences in which cooperative learning made things worse, so there is little danger in using it. It seems, however, that unless serious steps are taken to avoid a host of difficulties, the full potential of this approach could be missed.

Acknowledging that students are often quite inexperienced in cooperative learning, most of our respondents provide them with instruction on working with groups. This is done in several ways, usually at the beginning of the course. Some instructors hand out written suggestions or distribute articles to read and comments from students from previous courses in which this method was used. Others give a short lecture or conduct "brainstorming" sessions. In many cases, formal input from the instructor occurs throughout the course. Just before an exam is a favorite time. In many cases, the instructor takes time as needed to comment on issues that have arisen or gives a brief pep talk. This can also be done informally as the instructor is wandering through the class while students are working in groups on the computer or on classroom tasks. The meetings with groups described below provide another opportunity to make specific suggestions.

In general terms, probably the most important thing to do in avoiding difficulties is to see that the students are "sold" on cooperative learning, that they come to believe that this approach will work for them. It is necessary to convince students that more will be accomplished in a more pleasant atmosphere by working in a group than by working individually.

This can be accomplished in several ways. It is crucial that the instructor be an enthusiastic advocate of working cooperatively. He or she should convey to the students an enthusiasm for this approach and a strong confidence in its effectiveness. The instructor should explain the reasons for using cooperative learning and indicate some of the theoretical basis for it. Every time the students have some success, no matter how large or small, towards which this pedagogical strategy may have contributed, the instructor should talk about what happened and emphasize its connection with cooperative learning.

In addition to establishing an atmosphere of enthusiasm about cooperative learning, there were a number of specific things which our respondents did to avoid or overcome difficulties. The most useful technique was to meet regularly with each group. A half hour spent with each group about every two weeks does not amount to an excess of office hours. Using the techniques of group dynamics, many of the difficulties discussed in the previous section can be overcome. Presumably, through observations of the class, looking at written work, reading student journals and talking with the students, the instructor comes to the meeting of the group with a lot of information about how things are going. Using this, it is possible to get the members of the group to talk openly about their difficulties.

We have found that in such discussions, an open, concerned and accepting instructor can influence students to go a long way towards changing their behavior. Frequently a solution to a problem can be found when the problem is discussed openly and students want to solve it. We have rarely seen examples of a group that decided it really wanted to find times to meet and could not do so. The group can take on the responsibility of convincing an uncooperative member to change, of helping individuals who are having difficulties, or of toning down a group member whose enthusiasm has been stifling others.

In these group meetings our respondents have taken several points of view. Some try to convey to the group that it is up to them to work out their problems — with help from the instructor. The instructor can make specific suggestions, but only if he or she is convinced that the group members are ready to hear them. It is better if the ideas come from the students. In this situation we believe that the traditional Socratic approach can be effective. One approach is for the instructor to present, or summarize, what has already been presented in the discussion, list the various options that have been identified, and then let the students solve the problem.

Another tool, already discussed earlier, that can be used for finding out about group difficulties and helping students to overcome them is the use of journals. Some of our respondents ask the students to keep a journal of their experiences in and thoughts about the course. These journals are collected regularly by the instructor, read and returned. Reading the journals can tell the instructor a great deal about what is going on, and writing suggestions in them before returning them can be a way to get students thinking about solutions.

In a surprisingly and comfortingly small number of cases, direct action can be called for. Most of our respondents require each member of the group to sign each submission (of homework, exams, etc.) to indicate her or his participation. The group has the option of not allowing one member to sign. The possibility of such an exclusion can be a deterrent, and actually doing it can get a student to change behavior quickly. The instructor can act in drastic manners also, for example by changing the membership of a group.

We note that these extreme measures are fairly rare. Almost all groups will, at one time or another, have some of the difficulties we have described. In the overwhelming majority of situations, however, the methods we have described are effective in overcoming them.

6.4 Chapter Summary

The point of this chapter is that, as a pedagogical strategy, cooperative learning is not a panacea, but a potential. Although it is unlikely that using cooperative learning could lead to a significant reduction in the quality of learning, it will only lead to major improvements if it is implemented in a productive manner.

As with any pedagogical strategy, some difficulties may be expected. Most of the respondents to our Survey on Cooperative Learning reported that, on the whole, most of the groups worked in productive ways most of the time. However, there are various nonproductive behaviors that individuals and groups can engage in. In groups where the non-productive behaviors outweigh productive behaviors so that very little learning is taking place, the group may be called *dysfunctional.*

In this chapter we discuss various ways in which a group may function, describe some of the difficulties that have been reported by the respondents to our survey, and offer some strategies for dealing with these difficulties.

In working on class assignments, each group tends to develop a preference for one of the following modes of operation:

- work together on every problem,

- divide into subgroups to work on the problems, then come together to compare results,

- work individually on nearly every problem, and get together as a group to compare results,

- use some mixture of these strategies.

Whatever mode of operation the group tends to prefer, it is important that the members of the group do get together to discuss and reflect on their results.

The difficulties that groups have may be connected with *modes of operation*, may arise from *organizational issues*, may be connected with *group dynamics*, may have to do with *individuals*, and may have to do with *influences from the rest of the world*. We discuss examples of each of these, and offer suggestions and strategies for dealing with them.

In order to enhance productive group behaviors, it is important for the instructor to be aware of how the various groups are functioning. Instructors can learn about group dynamics by observing in the classroom and the computer laboratory, talking informally with students, reading student journals, and meeting formally with each group.

The value that students perceive the instructor places on cooperative work, both by example and by personal enthusiasm, can go a long way toward solving problems as they develop.

Chapter 7

What Are the Reactions of Students and Instructors?

Both students and instructors have much to say about their experiences with cooperative learning groups. We have collected students' comments written in course surveys, course evaluations, and journals as well as instructors' statements written as part of their response to the Survey on Cooperative Learning. These written reactions to individual experiences have confirmed our belief in the inherent benefits of cooperative learning, and they have helped us to refine our implementations of cooperative learning. In our own classes, students' written communications have allowed us to monitor the progress of the groups and to carry on a dialogue with individual students.

Students wrote about all aspects of their cooperative learning experiences. They often reported on the activities of the group — when the group met, how the work was accomplished, and how each group member contributed to the assignment. In addition, students frequently expressed their opinions and described their feelings, which often changed from week to week. We have seen great variation in the opinions of students, as shown by two comments illustrating the extremes:

> I thank God that I work in a group and we can all help each other.
>
> *Student journal, Calculus I*

> I am completely frustrated by my group. ...In class today I was absolutely in shock when I learned that the assignment due Friday was handed in late and incomplete.
>
> *Student journal, Calculus II*

Instructors responding to the Survey on Cooperative Learning addressed the same issues as the students, and in many cases, the instructors made observations similar to those of the students.

7.1   On learning and doing mathematics in groups

Students frequently wrote about their experiences learning mathematics and doing their assignments together. Many found that they learned well from each other and, in fact, some students claimed that their peers could explain ideas better than their instructors. Students recognized that they learn by explaining mathematics to their groups in their own terms.

Students pointed out the differences between their past experiences of working in isolation and their new experiences in their groups. They wrote about abandoning a problem once they were stuck while working alone rather than turning to group members for consultation on the difficult parts. They realized that a group discussion often generates a variety of ways of solving a problem and that a group solution may be better than any individual could have produced. They found that they persisted much longer on one problem when they were working together. Students wrote about checking each others' work to increase their understanding as well as to correct mistakes. They knew that they were learning to talk about mathematics, even debate mathematical positions.

Perhaps of most interest to instructors considering cooperative learning are the conclusions regarding the learning of mathematics reached by the respondents to the Survey on Cooperative Learning. Respondents wrote of the many advantages of cooperative learning that they found in their own classes. Some observed that these students developed a deeper understanding of the mathematical concepts and that the students learned to talk about mathematics. Others found that the learning seemed more uniform throughout the class, fewer questions went unanswered, and that the weaker students did not fall behind as quickly. On the other hand, the stronger students clarified their own mathematical ideas in the role they assumed as tutors within their groups. Some wrote that problem-solving skills were improved and that students, through their discussions with others in their groups, realized that there are different approaches to solving a problem.

7.2   On changing attitudes

Students wrote about changes in their attitudes towards mathematics fostered by their cooperative learning experiences. Some students explained how they had gained confidence in their mathematical abilities in the less threatening atmosphere of their small groups. They expressed their pleasure, enjoyment, or fun in solving problems in groups, and the interest promulgated by working in groups. One student wrote about the frustration of working for an hour with his group on one problem and then the relief when the problem finally was solved. Others mentioned the comfort of knowing that there were other students they could consult whenever they wanted to talk about a mathematical problem.

Respondents to the Survey on Cooperative Learning also wrote about the confidence that students gained by working in their groups. They observed that some students acquired a greater sense of self worth and a belief that they could solve difficult problems. Some mentioned that certain students became aware of how much mathematics they and their group members already knew and that, for the first time, students learned that their peers often have similar difficulties in mathematics. Some instructors found that their students developed a more positive attitude towards both mathematics and computers. They noticed a change from a competitive spirit to one of cooperation and mutual support within the class. Furthermore, on the question of the Survey on Cooperative Learning that asked how the attrition rate is affected by cooperative learning, about half of the respondents chose the answer that fewer students drop the course, and most of the others thought that cooperative learning has no effect on attrition.

7.3   On social skills and socializing

A few students noted that there are social gains in addition to academic gains gleaned by working cooperatively. These students wrote about learning to work with others as preparation for real life, where much work is accomplished by groups or teams of people assigned to a project. They discussed the friendships they formed with a diverse group of students as they learned to work together on their mathematical assignments.

Many of the respondents to the Survey on Cooperative Learning wrote about the development of social skills promoted by the group work. They found that their students were learning to work with others — to be considerate of the needs of others, to listen to others, to assist each other tactfully, to express appreciation for the help of another student, to take responsibility for part of a group project, and to work at communicating effectively. Instructors mentioned that some students had managed to work productively with others they did not much like and that they had learned to appreciate the different skills that their group members contributed to a project. They reported that some of the quieter students had learned to converse with their group members and to insist on their rights within the group. Students gained experience with addressing any difficulties with the dynamics of the group. The instructors observed that students within a group often became good friends as they learned the cooperative approach to academic work.

7.4   On difficulties and responsibilities

Students and instructors wrote about the difficulties that were confronted by some cooperative learning groups. Each of the possible difficulties, which are described in Section 6.2, has appeared in the student journals and on course surveys as well as in the comments written by the respondents to the Survey on Cooperative Learning. In fact, a student's articulation of the problem often was the first step in solving the problem. In reading journal entries of different members of the same group, the instructor frequently learned about a problem from different perspectives. For example, in one calculus class, the most able student in a group complained that she did all the work, and another student in the same group remarked that he received no feedback on his work because she always rewrote it.

Subsequent journal entries often traced the gradual elimination of a difficulty within a group. Some students responded to a conference with the instructor with a new awareness of their own contribution to a situation. In one case, a strong student, who had been unwilling to let the others in his group participate in any meaningful way, wrote about changing his expectations of the others in his group.

7.5   On group grades

Students wrote about their concern over the fairness of group grades. In particular, strong students worried that their individual grades might be adversely affected by their groups' work. Some of these strong students were especially protective of their high grades, and they didn't want their own work to be counted toward other students' grades. Typical remarks are:

> I don't want his laziness to destroy my grade like it almost did last semester.

> *Student journal, Calculus II*

> Why give someone somebody else's grade?

> *Student journal, Calculus I*

> Certain members of my group were happy with B- work. It pulled their grade up. They were uncooperative. It pulled mine down and they didn't care.

> *Course evaluation, sophomore Discrete Mathematics*

On the other hand, some students described the decreased pressure and tension within the course because their grade had a group component. They knew that the group work improved their grades in addition to helping them learn mathematics. One student wrote:

> When taking a test one person may not know everything but the others can contribute. Not only does this help your grade but it will help you learn the things you don't already know.

> *Student survey, Calculus II*

The respondents to the Survey on Cooperative Learning wrote that they would continue to use a group grade as part of the evaluation of the individual students within their classes.

7.6 On the whole semester

A few concluding quotations from students' writings near the completion of their courses further illustrate the variety of ideas and feelings of students participating in cooperative learning groups. Occasionally there is a group that never does overcome its difficulties, and some of the members, usually the most talented members, are not supportive of cooperative learning throughout the course. Their remarks show their inflexibility in fully participating in a method of learning different from one that has worked well for them in the past.

> To sum up my feelings, forced group activities aren't beneficial and in some cases are extremely unfair. I agree that peer help sessions do work, but not when you are forced to work with people who don't care about learning the material YET benefit from somebody else's hard work and dedication. ...

> *Student journal, Calculus II*

> Unfortunately I don't think that the educators know what it is like to work in a group. They may have experience teaching such a class but it is very different when you are being taught in such a class,

especially at this age. As I said, I did benefit from this course but not as much as I thought I would. Maybe I am just too used to traditional methods.

Student journal, Calculus II

On the other hand, by the end of the course, many students wrote enthusiastically about their cooperative learning groups. A few typical quotations from students follow:

As for working in groups I feel it is a good idea. First because it would be too much work for just one person to do. Second, it is good because if I get stumped on one of the problems I always know that I have three other people to call and ask for help. I think that is the best thing about working in groups because I know I called my group members many a time.

Student journal, Calculus II

We worked in groups which helped spark one another. It seemed as if each person put in their own way of doing mathematical problems.

Student survey, Calculus II

I have felt the frustration with complex problems which have discouraged me to go on with math but when that point of frustration hits it brings a good relief when you can turn to another group member and get some insight.

Student in Calculus I

The main thing I gained was excitement. It allowed me to get to know classmates better, and it also allowed me to enjoy doing math by doing it with someone rather than by myself.

Student survey, Calculus I

There was one question on the Survey on Cooperative Learning that summarized the respondents' evaluation of cooperative learning. When asked to choose an overall description of their cooperative learning experiences, no instructor chose the answer that cooperative learning was too problematic to continue. About sixty percent of the instructors answered that the cooperative learning method was generally successful, while the others chose the answer that there were some problems, but some positive student gains.

7.7 Chapter Summary

Students' written descriptions of their cooperative learning experiences and the comments of the respondents to the Survey on Cooperative Learning address many aspects of cooperative learning in undergraduate mathematics courses. The students and their instructors have written about learning and doing mathematics in groups, changing attitudes towards mathematics, affecting social skills, recognizing and addressing problems within a group, and grading. They have also described their opinions on cooperative learning at the end of a mathematics course.

What do the students and instructors say about cooperative learning? We know that instructors who have tried cooperative learning say that they are willing to continue with this method of instruction, and that many students recognize the gains made by working in groups. The support of a group seems to foster lower attrition rates in some classes as well as improved attitudes towards mathematics. Group work helps to prepare the students for the real world, where many projects will be accomplished by a team effort without choice of colleagues. Most important of all, despite the difficulties that must be overcome by some groups, mathematics is being learned in a social setting that encourages discussion of the various aspects of a mathematical concept and expression of different ways to solve a problem.

Chapter 8

What Are Other People Doing?

In the previous seven chapters we describe a model of cooperative learning that is being implemented in a variety of mathematics classrooms at both the college and the high school levels. In this chapter we survey a small sample of other forms that cooperative learning may take in mathematics classrooms. It is not our intention to present a comprehensive, critical review of the field, but rather to give an overview to inform practitioners about a few more methods of cooperative learning that could be implemented in mathematics classrooms. Eight major techniques that have been investigated in secondary and undergraduate mathematics classrooms are summarized and discussed in this survey. Each assumes a traditional classroom of one instructor and many students organized into heterogeneous ability groups of four to five students working together to learn the material. Differences among the techniques are in the extent to which cooperative learning is accountable for individual achievement versus group productivity. References for further, more specific, information are given for each method included in this survey. We hope that this chapter, together with the bibliography, will give the reader some idea of what is available in the literature on cooperative learning.

8.1   Student teams-achievement divisions

Since this is the simplest method, it is recommended for instructors who are beginners in using cooperative learning methods in their classrooms. The major components of the method are class presentations, quizzes, individual improvement scores, and team recognition.

The instructor introduces the new unit in a class presentation. The difference from the traditional lecture presentation is that the instructor focuses clearly on the unit on which students will take quizzes. After the instructor presents the unit, the students are divided into teams to work on the worksheets or some other material. Students are assigned to four-member heterogeneous teams based on performance level, gender, and ethnicity. Some of the rules that should be followed while working in groups are:

- It is the students' responsibility to make sure their teammates have learned the material. The instructor may ask all the group members to sign at the bottom of their worksheet, indicating that everyone has mastered the material.

- The team cannot finish studying until all teammates have mastered the material.

- Students should ask teammates for help before asking the instructor. To assure this, all the students in a group are to raise their hands when they have a question no one in the group can answer.

After the material has been mastered, all students take quizzes or do worksheets individually, and they get the scores for their individual performances. The scores are compared to their own past averages, and points are awarded based on the degree to which students differ from their earlier performances. Those individual-improvement points are assigned to their groups to form a group score. Groups that meet certain criteria may earn certificates or some other reward (for example, a listing in a newsletter). This cooperative learning method has been used in mathematics classrooms from elementary through college level. It is especially recommended when teaching mathematical and computational skills. Instructor-prepared materials are not difficult to make; all that are necessary are a worksheet, answer key, and quiz for each unit (Slavin, 1986). Below are some ideas for the classroom activities suggested by Slavin (1990d).

Teaching: The instructor presents the lesson. It is usually a good idea to plan more than one day per unit. The main idea of the lesson should be clearly stated. The focus is on meaning, rather than memorization. Student comprehension is assessed by asking a lot of questions, and calling on the students at random. Manipulatives and other helpful tools are used to demonstrate concepts or skills.

Team study: Students work in teams on the worksheets or other materials. There could be one or two worksheets per group. The teammates are to help each other in mastering the material. The instructor should insist that the group members explain their answers to one another, not just compare their answers. While the students are working in their groups, the instructor should circulate among the groups, observing and listening to the students, and answering their questions only when none of the group members is able to give an answer.

Test: Students take individual quizzes that require about half a class period. While taking quizzes, students should not talk to each other. They should get their quizzes back in the next class period.

Team recognition: A computation of the individual improvement score is based on each student's base score. After the individual and team scores are computed, the highest team scores are announced, and appropriate team awards are given.

8.2 Teams-games-tournaments

A distinguishing characteristic of this method is that the students, as representatives of their teams, compete with students of other teams whose academic performance in the past has been like their own.

The new unit usually begins with the instructor's presentation followed by the students' work in heterogeneous four- or five-member groups on the worksheets for the lesson. Once a week the students play academic games with students from other teams who achieve at a similar level. The games are structured and conducted in the following way:

1. Students are assigned to triads of comparable academic achievers from three different groups.

2. About 30 content-related game questions are prepared and listed on a separate sheet of paper. The same questions are written on separate numbered cards. A student pulls a card and tries to answer the question. At the end of the competition, lasting about half of a class period, the students in each triad score their correct answers and make a rank order of the students.

3. Students return to their cooperative groups and derive their group score by adding the individual scores of all their members.

4. The highest scoring teams are publicly recognized and students' grades are based on individual performance (DeVries and Edwards, 1974).

8.3   Team-assisted individualization

Although this method is specifically designed to teach mathematics in the lower grades (grades three to six), it is also used as remedial instruction in secondary schools. It shares with the previous two methods the idea of heterogeneous groups and certificates for high performing groups.

The students work in groups on their individual worksheets. They ask each other for help, suggestions, and corrections. Students check each other's work against the answer sheet, and they help each other with problems. Basically, each student proceeds at her or his own rate.

Meanwhile the instructor may circulate among the groups, presenting new material to each group that has mastered the previous material.

At the completion of the whole unit, each student takes an individual test that is scored by student monitors. Since the scoring is put into students' hands, the instructor has time to present lessons to groups composed of students who are at the same level (Slavin, 1990c).

At the end of each week, a group score is calculated by counting the number of units completed by all group members and the number of final tests passed, and adding some points for homework or perfect papers. The groups that achieve a criterion score are rewarded.

In developing classroom materials to use with this method, the instructor starts with lesson guides that suggest methods of presentation, manipulatives, demonstrations, and computer investigations. Worksheets and answer keys need to be prepared.

8.4   Jigsaw

The students are divided into five- to six-member heterogeneous teams (based on ability, gender, and ethnicity) to work on an academic task. The academic task is broken into sections and every member of the group is supposed to study one section. Then members of different groups who have studied the same section meet together to discuss their sections. Afterwards, they return to their group and teach other members about their sections. Students are assigned a grade based on their individual performance on a quiz that covers the whole unit. A very important point in this method is interdependence: every student depends on others when learning the whole unit. Some modifications of the Jigsaw method have been developed at Johns Hopkins University (Aronson et al., 1978).

The Jigsaw method is very flexible. Various modifications are possible that keep the basic model but change some details. For example, a lesson can be broken into as many units as there are groups in the classroom. Then each can be further subdivided into five or six topics that make sense by themselves. The students are assigned to groups and given the unit to be studied. They decide how to split the topics among themselves. Each team prepares a report on its unit and presents the report to the class.

8.5   Learning together

The students are grouped heterogeneously with respect to gender, ability, and ethnic and cultural background. These may be formal learning groups (that stay together until the task has been mastered by all members of the group), informal groups (formed, for example, by pairing students who are sitting together), or base groups (long-term groups that last for a semester, a year, or even for several years). Students in informal groups may check each other's answers on a worksheet, or a student may check to see if another nearby student understands a concept. Students in base groups provide each other help needed to complete the assignment, and they give each other support and encouragement over a time period that goes beyond particular assignments or units of study (Johnson and Johnson, 1987c).

This method is appropriate for investigation of new material or application of material that has already been studied. The students work together following rules of cooperative learning that have been established in their classroom. They turn in a single assignment as the result of the group effort. Individual scores are based on the group product.

8.6   Group investigation

The whole class is divided into groups of three to five students. Groups can be formed using various criteria: friendship, preference for particular topic, or ethnic heterogeneity. Groups choose their strategy of study, and each student in the group chooses a subtopic to study. Then each member reports to the group the result of her or his study, and the group discusses the material in order to prepare a group report for the whole class. This method was developed by Shlomo and Yeal Sharan at the University of Tel-Aviv (Sharan and Hertz-Lazarowitz, 1980).

Instructors who use this method need to be aware of the background and abilities of the students, and they must know how much time they want to spend on each unit. The role of the instructor is to be resource person and facilitator. The problem or unit is presented to the students, and the groups decide how they will split the work. Each group should work on a topic that is part of the whole class unit. Students are responsible for organizing and working in their groups. They may choose to work together or individually on subtopics. At the end of the unit, all the groups organize and present their reports as a part of a whole class lesson. It is beneficial if all students are included in the final class report.

8.7 Coop-coop

This method is similar to the group investigation method. The class is divided into heterogeneous groups with respect to ability, ethnicity, and gender. As in group investigation, the material to be learned is divided into subunits so that each group is responsible for a subunit. Then, each group decides how to further divide its subunit into smaller parts so that each group member individually studies one part. After completing their work, all group members present their topics to the group. Then, the group prepares a report on its subunit and presents it to the whole class. Individual learning is evaluated through quizzes, which may be either oral or written (Kagan, 1985).

This method is suitable for group or class projects. It is very similar to the Jigsaw method. The unit to be studied is broken into as many individual topics as there are student teams. The students in each team break down their topic into smaller parts, and decide which member will research each part. Sufficient resources for research and student investigation should be available.

Depending on the topic, it may take one week or more for students to complete their research. After the completion of individual research, each member reports on her or his topic to the rest of the group. The group discusses the complete subunit, suggests improvements, and writes a report or prepares a presentation for the class. At the end, students take individual quizzes on each topic.

8.8 Small group laboratory approach

The groups are formed by "counting off"; that is, the class of n students is divided into groups of four by letting students count up to $\frac{n}{4}$. Students with the same numbers form a group. For example, in a class with 32 students there would be eight groups. The students would count off 1, 2, 3, 4, 5, 6, 7, 8, 1, 2, ..., 8, 1, 2, ... Then the 1's would form a group, 2's would make another group, and so on. In case of serious problems between group members, the groups could be reorganized. Otherwise, the group would work together for five to ten weeks. A preferable class size for the group work is 24, up to a maximum of 32 students.

Non-mathematical questions or topics are suggested to start the class. For example, students may be asked to share with others in their group a "memory of reading a good book," or "something you like about yourself," or "something you are interested in." These class-starters serve as a positive stimuli and help to establish good relationships among the students and instructor. The instructor may make a list of "Remarks on Learning Mathematics in Small Groups" in which expectations of behavior and performance of group members are stated explicitly. The students are reminded of one or two of these guidelines each week.

In developing a lesson plan, it is useful for the instructor to first think about the mathematical concepts that the students are studying and any prerequisite skills that may be needed. Then the instructor lists mathematical experiences and gathers manipulative materials that could be used to help students understand these concepts. Finally, class activities are devised and instructions are prepared that will enable the students in each small group to work through these activities and learn the concept. See Weissglass (1985, 1990) for some examples of such lesson plans.

This method has been developed and implemented by Weissglass (1990) for teaching college-level students, mathematics majors and pre-service elementary teachers.

8.9 The small-group discovery method

This is a very flexible method in terms of group formation, size and duration. It can be used as a total instructional method or in combination with other methods. The groups may be formed by the instructor as either heterogeneous groups (with respect to gender, achievement, race) or homogeneous groups based on students' mathematical achievement; or the groups may be formed by having the students choose their own group members. Group size may vary from three to five students, and the groups can last from one class period to one semester. Each group discusses mathematical ideas and solves problems cooperatively. The emphasis is placed upon the discovery of new ideas by the students. At the beginning, these ideas are very informal in character but change as the students become more experienced. While students work in their groups during the class period, the instructor walks around checking the group work and providing assistance where needed. During discussion with the whole class, the instructor serves as discussion moderator and usually summarizes what the students have found. The instructor provides guidelines for cooperative learning and may incorporate them into the instructions for the classroom activities. A variety of grading schemes could be used with this method, including homework, classwork, quizzes, tests (group tests, take-home tests, in-class tests), and projects (Davidson, 1990a).

This method is very suitable when students work in computer or calculator laboratories. The instructor may direct each group to turn in one assignment as the group product. Since emphasis is on discovery, the grading on such assignments should not be based solely on the "correctness" of the answer, but rather include some measure of progress toward the concepts to be discovered.

8.10   Chapter Summary

In this chapter we have offered a limited survey of other forms of cooperative learning that are being used in mathematics classrooms. The methods surveyed include both formal and informal methods of group formation, and class activities that are more and less structured.

Groups may be heterogeneous with respect to criteria such as gender, race, and ethnicity, or homogeneous with respect to cultural and ethnic background or mathematical ability and achievement. Groups may be formed by a random mechanism such as "counting off," or by pairs of students who are sitting near each other.

Informal groups may stay together for a single class activity. Formal groups may stay together for a longer assignment or class activity that may take several class periods; groups may stay together for the duration of a unit of study. The unit may be divided into subunits, with each group member responsible for studying one of the subunits individually and reporting back to their group. Individuals from different groups who are working on the same subunit may meet together to clarify their ideas around particular concepts before presenting the material in their own groups. In some methods, each group makes a report to the whole class.

Longer-term base groups that stay together for a semester, a year, or even several years provide support and encouragement for group members in addition to help on particular assignments.

Various methods are used to assess and reward individual and group achievement. Group projects or reports may be followed by individual quizzes. Groups may be assigned points based on achievement or improvement in the scores of individual members. Public recognition may be given in the form of certificates; special achievements by groups may be mentioned in a newsletter. Individual grades may be based on individual performance on quizzes and tests, and may also reflect group achievements. Overall there is an effort to reward and encourage group performance while also requiring individual accountability.

The various methods surveyed offer different combinations of cooperation and interdependence among group members, competition between groups, rewards for individual and/or group achievements, and assessment of individual learning. Each of the methods surveyed includes, to varying degrees, some of the components that the authors of this monograph consider essential to cooperative learning (see Chapter 1). As was pointed out in Chapter 2, assessments of the relative effectiveness of "cooperative learning" can be difficult to make unless careful attention is paid to the underlying assumptions about cooperative learning, and the elements of cooperative learning that are present in the methods being compared. We leave it to the reader to reflect on how well each method surveyed in this chapter encourages and challenges students to learn new mathematical concepts.

Appendix A

A Survey on Cooperative Learning

This Survey on Cooperative Learning was developed by the authors in the Summer of 1992, and sent out to about eighty of our colleagues the following Fall. We received forty-two responses.

We are interested in learning the extent and variations of cooperative learning methods employed in collegiate mathematics courses. Our aim is to develop a practical guide to cooperative learning in collegiate mathematics based on our own experience and the experience of others. By cooperative learning we mean working and studying in formal groups for an extended period of time.

Please answer the following questions. In addition, we would appreciate your comments on any aspect of your experiences with cooperative learning and any other material, such as student comments in journals and course evaluations. Thank you very much for your participation in this survey.

1. In which courses have you used cooperative learning?

   (a) Precalculus

   (b) Calculus I

   (c) Calculus II

   (d) Calculus III

   (e) Discrete Mathematics

   (f) Linear Algebra

   (g) Abstract Algebra

   (h) Other (Please specify.) ....

2. What size groups have you used most commonly? Please comment on why you used this size.

   (a) 2 students

   (b) 3 students

   (c) 4 students

   (d) 5 students

   (e) 6 students

   (f) Other (Please specify.) ....

3. What size group do you think works best? Please comment on why you think this works best.

   (a) 2 students

   (b) 3 students

   (c) 4 students

   (d) 5 students

   (e) 6 students

   (f) Other (Please specify.) ....

4. How were the groups formed? Circle one.

   (a) Students formed their own groups.

   (b) I, the instructor, assigned students to the groups.

   (c) Other (Please specify.) ....

   Please describe further how the groups were formed.

5. If you assigned students to the groups, what criteria did you use? Circle all that apply.

   (a) Math SAT

   (b) Verbal SAT

   (c) Average grades in all previous courses

   (d) Average grades in previous mathematics courses

   (e) Previous mathematics courses

   (f) Other particular knowledge, such as computer expertise
       (Please specify.)

    (g) Gender

    (h) Placement test at entry

    (i) Race

    (j) Nationality

    (k) Age

    (l) Common free time for group meetings

    (m) Other (Please specify.) ....

6. How did you use the criteria? Circle one.

    (a) For homogeneous grouping with respect to all criteria used

    (b) For heterogeneous groupings with respect to all criteria (except common free time for group meetings)

    (c) Other (Please specify.) ....

7. How long are the groups usually maintained? Please comment on why you used that length of time.

    (a) Less than half a semester

    (b) For about half a semester

    (c) For about a whole semester

    (d) For about two semesters

    (e) For a major project

    (f) Other (Please specify.) ....

8. For which activities have you used cooperative learning? Circle all that apply. Please comment on the relative effectiveness of the different types of cooperative learning activities and/or on any specific details.

    (a) Problem-solving in class with class discussion immediately following

    (b) Problem-solving in class on longer, harder problems

    (c) Working during supervised computer laboratories

    (d) Homework assignments

    (e) Tests and/or examinations

    (f) Other (Please specify.) ....

9. If you used group testing, please answer the following.

   (a) I usually give ...... group tests out of a total of ...... tests during the semester.

   (b) Have you ever given a group final examination? ...... If so, please comment on the students' reaction.

   (c) What percentage of the total test and examination grade was earned through group tests?

10. What percentage of a student's grade is earned through group activities?

11. Do you give students instruction in cooperative learning?

    If so, when do you give instruction? Circle all that apply.

    (a) Initially. Please describe the instruction.

    (b) Throughout the time. Please describe.

12. How do you monitor the groups? Circle all that apply.

    (a) Student journals

    (b) Questionnaires

    (c) Informal observations

    (d) Formal meetings with me

    (e) Other (Please specify.)

13. What non-productive behavior did you encounter with your groups? Circle all that apply.

    (a) One student did not do a fair share of the work.

    (b) One student dominated discussions.

    (c) One or more students did not contribute to discussions.

    (d) Some students did not attend planned group meetings.

    (e) Students could not find a common time to meet.

    (f) Students did not want to meet outside of class.

    (g) A group split into subgroups that had little interaction.

    (h) Students divided the group work and then had little interaction.

    (i) Other (Please specify.) ....

14. Please describe the problems of any dysfunctional groups or individuals.

15. What did you try in attempting to handle any problems in cooperative learning? Circle all that apply.

    (a) I usually ignored the problems.

    (b) I met with any such group as a whole.

    (c) I met with the members of any such group individually.

    (d) I wrote suggestions in student journals.

    (e) Other (Please specify.) ....

    How effective was your response in each case?

16. When you communicate with students about their problems with cooperative learning, what is your usual point of view? Circle one.

    (a) It is up to the students to work out their own problems.

    (b) I make specific suggestions in an effort to improve the functioning of the group.

    (c) I present options and then let the students solve the problems.

    (d) Other (Please specify.) ....

    Please comment on your experiences with communicating with students about any problems they encounter with cooperative learning.

17. Was technology used by the groups in your class? ......If yes, circle all that apply.

    (a) Mathematical programming language (e.g., *ISETL*)

    (b) Computer algebra system (e.g., *Derive, Maple, Mathematica*)

    (c) Software accompanying a textbook

    (d) Special software (Please specify.) ....

18. What was the effect of the use of technology?

    (a) Interfered with group dynamics

    (b) Improved group dynamics

    (c) No effect

    (d) Different effects on different groups

19. How would you describe your cooperative learning experiences? Circle one.

    (a) Generally too problematic to continue cooperative learning

    (b) Some problems, but some positive student gains

    (c) Generally successful

20. How is the attrition rate affected by cooperative learning? Circle one.

    (a) Fewer students drop courses in which cooperative learning is a major feature.

    (b) More students drop such a course.

    (c) Cooperative learning seems to have no effect on attrition.

21. How did most students react to cooperative learning?

    (a) Very positively

    (b) Positively

    (c) Neutral

    (d) Negatively

Please send copies of any students' written comments that you are willing to share.

22. What do you think that your students have gained from their cooperative learning experiences?

23. Please make any other comments that you think might be helpful.

Again, thank you very much for your time and effort in answering these questions.

Appendix B

Responses to the Survey

For the convenience of the reader, the survey questions are repeated here along with the tallies of responses and comments from our respondents. Comments are reported here almost verbatim, with minor editing for spelling, etc.

Forty-two colleagues answered our survey. When a respondent chose several answers, we counted each answer.

1. *In which courses have you used cooperative learning?*

course	number of responses	class size
Precalculus	8	17 – 40
Calculus I	35	7 – 65
Calculus II	20	7 – 60
Calculus III	4	5 – 7
Discrete Mathematics	12	4 – 25
Abstract Algebra	6	8 – 14

Other (Please specify.) (32 responses)

course	class size
Finite Mathematics	16
Mathematics for Liberal Arts	26
Excursions into Mathematics	30
Course for Prospective Elementary Education Teachers	
Mathematics for Elementary Education Majors	9 – 32
College Algebra	48
(Continued on next page.)	

course	class size
Differential Equations	20
Linear Algebra	22
Linear Algebra & Discrete Mathematics (honors)	12
Non-Euclidean Geometry	22
Numerical Analysis	3 – 17
Statistics	26 – 29
Probability	
Elementary Statistics	25 – 30
Introduction to Statistics	30
Probability & Statistics	10
Data Analysis	20 – 25
SAGE 101–102 (honors)	16
Senior Seminar	7 – 15
Computer Literacy	20
Computer Science I	21
Pascal Programming	20 – 25
Computer Programming	15
Systems Analysis	32

2. *What size groups have you used most commonly? Please comment on why you used this size.*

   (a) *2 students* (2 responses)
      - no reason for using 2, have changed to 4 recently
      - lab is equipped with 12 machines, 2 students per machine

   (b) *3 students* (16 responses)
      - was best for class size
      - put 3 because of computer availability, but prefer 4
      - it is easier for 3 to find common time than 4 or more
      - 3 works well in small classes
      - 3 is most convenient
      - class size is 12 so 3 works best
      - 3 because odd numbers are preferred
      - in the computer lab, 3 students on 2 computers works well
      - it is easier to divide class into categories when 3 are in a group
      - convenience and balance according to the number of computers available
      - 3 because you need a tie-breaker

(c) *4 students* (26 responses)

- literature indicates 4 students are most effective (3 responses)
- they can split into 2 groups of 2 (2 responses)
- 4 is ideal — nice subdivision
- 4 to balance the number of boys and girls
- groups of 4 avoids 2 against 1
- 4 is optimal for completing homework assignments on time
- 4 in a group is able to sustain itself if one is absent
- 4 encourages working in pairs
- 4 is the most appropriate size to complete homework
- fit best with the number of computers
- suggested by the majority of those who tried it

(d) *5 or more students* (no responses)

(e) *Other (Please specify.)* ....

3. *What size group do you think works best? Please comment on why you think this works best.*

(a) *2 students* (1 response)

(b) *3 students* (12 responses)

- small class size
- easiest to arrange schedules
- ensures adequate exchange of ideas
- $\frac{2}{3}$ majority opinion worked almost all the time and was easy to reach
- would need more experience to separate (b) and (c)

(c) *4 students* (27 responses)

- can break down further into pairs
- group continues when one is out
- not too large — not too small
- better interaction
- to encourage efficient work habits, reflective thinking, and heterogeneous composition
- provides adequate mix of individual needs and resources
- 4 gives a stable configuration

(d) *5 students* (1 response)

4. *How were the groups formed? Circle one. Please describe further how the groups were formed.*

  (a) *Students formed their own groups.* (11 responses)

    - Students were given 2 days to choose partners who are likely to understand teamwork.
    - I put non-workers in their own group.
    - The students formed groups of 2, and then I combined them for groups of 4.
    - Students picked by availability of free time.

  (b) *I, the instructor, assigned students to the groups.* (30 responses)

    - wanted partners in the same dorm
    - balanced groups with math grades, computer background, gender
    - I wanted at least 2 women per group or no women per group, and at least one common free period per group.
    - I sorted by criteria.
    - I sort after the first test.
    - by ability, computer experience, math courses completed, and gender
    - Scheduling on a commuter campus is all important.
    - I have a questionnaire.
    - balanced talent as shown on questionnaire, one sharp student and one who needs help

  (c) *Other (Please specify.)*

    Instructor guided process, but students chose the groups (1 response)

5. *If you assigned students to the groups, what criteria did you use? Circle all that apply.*

  (a) *Math SAT* (12 responses)

  (b) *Verbal SAT* (4 responses)

  (c) *Average grades in all previous courses* (4 responses)

  (d) *Average grades in previous mathematics courses* (13 responses)

  (e) *Previous mathematics courses* (21 responses)

  (f) *Computer expertise* (27 responses)

  (g) *Other particular knowledge* (9 responses)

  (h) *Placement test at entry* (3 responses)

(i) *Gender* (24 responses)

(j) *Race* (4 responses)

(k) *Nationality* (no responses)

(l) *Age* (4 responses)

(m) *Common free time for group meetings* (13 responses)

(n) *Other (Please specify.)* (5 responses)
- student's expressed preference
- attitudes towards math (2 responses)
- residence hall (2 responses)
- try to have "computer person" and "math person" in each group
- logic test
- typing skill
- ownership of graphing calculator

6. *How did you use the criteria? Circle one.*

   (a) *For homogeneous grouping with respect to all criteria used* (3 responses)

   (b) *For heterogenous groupings with respect to all criteria used (except common free time for group meetings)* (29 responses)

   (c) *Other (Please specify.)* (3 responses)
   - Each group should have someone "smart."
   - Older (35+) students were paired and then placed with a younger group.

7. *How long are the groups usually maintained? Please comment on why you used that length of time.*

   (a) *Less than half a semester* (3 responses)
   - first grouping is hit or miss
   - reformed groups several times
   - students learn to work with variety of personalities

   (b) *For about half a semester* (2 responses)
   - best for regrouping

   (c) *For about a whole semester* (28 responses)
   - good working relationship
   - resubmission policies — students keep submitting until work is correct

- tried not to switch except as a last resort
- seems to work
- takes a while for students to feel comfortable with each other
- suggested by Dubinsky & Schwingendorf
- difficult to rearrange the groups
- course length — some degree of stability
- to model the work environment
- to develop as a group identity
- effectiveness of work/resistance to changing group dynamics typically increase with time
- it takes 3 weeks to work together — not reasonable to reform in middle of semester
- to make them less competitive with their group and to give them the impression that they "sink or swim together"
- must revise the groups in the second semester due to F's and transfers
- I wanted to spread the ones who were socially successful to influence and improve the other groups.
- So students will know from the outset that they need to make an effort to participate in their group.

(d) *For about two semesters* (3 responses)

- The more students work together, the better they utilize their strengths and weaknesses.
- students' choice
- by end of semester the students are getting "reasonable" grades

(e) *For a major project* (1 response)

- groups are changed approximately 3 times during the semester

(f) *Other (Please specify.)* (3 responses)

- used all the options at least once
- the goal of the project drives the decision

8. *For which activities have you used cooperative learning? Circle all that apply.*

(a) *Problem-solving in class with class discussion immediately following* (41 responses)

(b) *Problem-solving in class on longer, harder problems* (25 responses)

(c) *Working during supervised computer laboratories* (35 responses)

(d) *Homework assignments* (38 responses)

(e) *Tests and/or examinations* (32 responses)

(f) *Other (Please specify.)* (5 responses)

- group portfolios in numerical analysis
- reading text and discussing
- projects related to materials to be presented in class
- specific projects in statistics
- short lecture / problem-solving

Please comment on the relative effectiveness of the different types of cooperative learning activities and/or on any specific details.

- The pre-planning and requirement outcomes are key.
- Each class is different and responds differently to each method.
- Using groups for homework assignments saves time.
- For quizzes I grade at random — one paper per group saves time.
- Cooperative groups work well for everything.
- The problems need to be hard enough that they want to ask each other.
- The elementary education majors have consistently been the least cooperative.
- combination of group and individual efforts on quizzes
- Student becomes coach rather than a passive receiver of information.
- Students enjoy problem solving in class, but I think the group problem solving in labs and outside class contributes more to their growth.
- "Problem solving in class with discussion immediately following" is least favorite activity.
- The computer labs seem to go well.
- Students actually talk mathematics. They think through problems together instead of simply memorizing procedures for finding solutions. Students read each other's work.

9. *If you used group testing, please answer the following.*

   (a) *I usually give ...... groups tests out of a total of ...... tests during the semester.*

Responses to this question were:
2/10, 3/11, 1/4, 4/10, 2/5, 1/3 (8 responses), 1/2 (2 responses), 2/4 (6 responses), 3/6, 4/8 (2 responses), 2/3 (2 responses), 4/5, 4/4 (each test had an individual section and a group section).

(b) *Have you ever given a group final examination? ...... If so, please comment on the students' reaction.*

Responses to this question were:
27 "no" and 4 "yes."

- No special reaction as it is distributed one or two weeks prior to being due, and in-class individual part of final focuses upon somewhat easier problems than does the "take home."

(c) *What percentage of the total test and examination grade was earned through group tests?*

Responses to this question were:
5 – 15%, 10 – 15%, 14%, 15%, 15 – 40%, 20%, 25% (4 responses), 25 – 40%, 30 – 40% (4 responses), 30 – 60%, 40% (5 responses), 40 – 50%, 50% (5 responses), 60% (2 responses).

10. *What percentage of a student's grade is earned through group activities?*

Responses to this question were:
5%, 14%, 10 – 30%, 12.5%, 20% (5 responses), 20 – 25%, 20 – 30%, 25% (3 responses), 25 – 65%, 30% (2 responses), 30 – 40% (2 responses), 30 – 60%, 35%, 40% (5 responses), 50% (9 responses), 50 – 65%, 50 – 70%, 60%, 66%, 67%, 90%.

11. *Do you give students instruction in cooperative learning?*

34 responded "yes," 10 responded "no." Comments were "very little," "it varies," "informal and individual."

If so, when do you give instruction? Circle all that apply.

(a) *Initially. Please describe the instruction.*

(20 responses)
handouts, articles, and short lecture; comments from previous students; "brainstorming" and "debriefing"; using a guide from *Key Curriculum Press*; motivate only; go through the steps to successful cooperative learning.
I talk about reasons for groups, for same groups throughout term, and some good "practices" to consider/use.

(b) *Throughout the time. Please describe.*

> (12 responses)
> before first exam; continued debriefing; comment on issues that arise; pep talks as needed; mini-lecture in class; during my "wanderings" and during one-on-one sessions in my office.

12. *How do you monitor the groups? Circle all that apply.*

(a) *Student journals* (16 responses)

(b) *Questionnaires* (12 responses)

(c) *Informal observations* (34 responses)

(d) *Formal meetings with me* (19 responses)

(e) *Other (Please specify.)* (5 responses)

- in-class work and papers submitted daily from the team
- performance, chats with individuals in class
- assignments handed in for grading
- During class time I wander around the room and drop in on various groups.
- formal meetings with student consultants

13. *What non-productive behavior did you encounter with your groups? Circle all that apply.*

(a) *One student did not do a fair share of the work.* (34 responses)

(b) *One student dominated discussions.* (22 responses)

(c) *One or more students did not contribute to discussions.* (31 responses)

(d) *Some students did not attend planned group meetings.* (24 responses)

(e) *Students could not find a common time to meet.* (21 responses)

(f) *Students did not want to meet outside of class.* (16 responses)

(g) *A group split into subgroups that had little interaction.* (15 responses)

(h) *Students divided the group work and then had little interaction.* (26 responses)

(i) *Other (Please specify.)* (2 responses)

- A few cases in which one student "promised" to be responsible for certain problems on a group assignment and did not carry through. In all these cases but one, the group took action to correct the situation.

- One student left the group to work individually because he lived far away.
- One student left school due to illness.
- One student abandoned one group for another.
- One student was overwhelmed with the amount of work.

14. *Please describe the problems of any dysfunctional groups or individuals.*

- put "non contributors" into their own group and they did the work
- had one group with "battle of the sexes," 2 boys and 2 girls
- inability to meet at a common time
- communication problems with foreign students
- Extremely good students felt held back by the group.
- Students with little self-confidence don't speak out.
- The biggest problem is students who refuse to do their share of the work.
- excuses; inconsistencies; low test scores; once group is dysfunctional, it tends to stay dysfunctional
- personality conflicts
- serious illness
- bad mismatch of abilities leading best students to abandon two weaker students to work with each other
- serious computer phobia
- Students fail to ask for help until the due date.
- One member of the group wants to prove to me that she knows it all. She sits with her partner, but for the most part works alone.
- One member of a group of four who can't seem to say to her partners "stop" or "slow down" — because of pride she refuses to admit that she's lost.
- chronic absenteeism by one member of the group
- one member was dominating/stifling with help
- reluctance of group to penalize those who did nothing

15. *What did you try in attempting to handle any problems in cooperative learning? Circle all that apply.*

    (a) *I usually ignored the problems.* (6 responses)

    (b) *I met with any such group as a whole.* (28 responses)

    (c) *I met with the members of any such group individually.* (31 responses)

    (d) *I wrote suggestions in student journals.* (9 responses)

    (e) *Other (Please specify.)* (4 responses)

    How effective was your response in each case?

    - fair, sometimes worked (7 responses)
    - I don't know. (2 responses)
    - good (9 responses)
    - very good (5 responses)
    - not very good (3 responses)
    - There was some awareness of responsibility.

16. *When you communicate with students about their problems with cooperative learning, what is your usual point of view? Circle one.*

    (a) *It is up to the students to work out their own problems.* (10 responses)

    (b) *I make specific suggestions in an effort to improve the functioning of the group.* (22 responses)

    (c) *I present options and then let the students solve the problems.* (22 responses)

    (d) *Other (Please specify.)* (4 responses)

    - If the problem is not solved, take specific action.
    - Brainstorming sessions between teacher and student.
    - Group can prevent students from signing "cover sheet" if they don't participate.
    - I asked them to write down the problems and I read them in class. It makes a difference.

Please comment on your experiences with communicating with students about any problems they encounter with cooperative learning.

- Most students indicated willingness to try to make adaptation, but very often do not follow through.

- There are students who have had calculus that have a difficult time accepting a different approach to learning. Sometimes we need a "truce" so that the student will accept the approach and give it some time before quitting the team.

- I observed a lack of concentration from the rest of the group while one member of the group was working on the computer.

- Good students think that it takes too much of their time.

- I try to keep the problem in a 3rd person scenario and focus on the importance of completing the *goal* of the group. Anytime we are involved in cooperative learning, we are asking students to change a long-standing paradigm. Patience, kindness, and student dignity are keys to success.

- I usually try to show the students how they have choices.

- Some students try to insist on a teacher-initiated solution.

- It depends on the problem.

- The students are generally aware of their shortcomings. Some take the leadership and redeem themselves. Others persist in excuses and fail the course.

- Usually quite reluctant to say something "bad" about another and tensions high at first — however, after a time "someone" will volunteer they haven't been holding up their end of bargain and progress starts at that point.

- Usually after a meeting with a group, I see improvement.

- I feel this is something I need the most help with myself. Cooperative learning was never really a part of my own educational experience.

- I try to create an environment where working as a team functions better than working individually.

- Most are willing to talk about it, but they always blame problems on others in the group. I have difficulty getting them to accept responsibility for problems.

- I have one student who is far above the other two in his group and he is concerned about their pulling down his grade. I had to convince him that this wasn't going to happen. Consequently he adjusted to the situation and things are going fine now.

- Some students simply cannot get along. Personality, learning style, motivation and expectations are more important than talent and experience.

- First I ask the students to describe to me how they think things are going. I then make specific suggestions, offering several options that I feel might help improve the functioning of the group.

17. *Was technology used by the groups in your class?*

   "yes" (41 responses), "no" (2 responses)

If yes, circle all that apply.

(a) *Mathematical programming language (e.g., ISETL)* (34 responses)

(b) *Computer algebra system (e.g., Derive, Maple, Mathematica)* (29 responses)

(c) *Software accompanying a textbook* (6 responses)

(d) *Special software (Please specify.)* (9 responses)

  - TI-81 calculators are required of all students
  - Fields and operators, version 3.0
  - graphics calculators
  - Mathematics Plotting Package, MPP (USNA)
  - MINITAB spreadsheets
  - Pascal, BASIC — These were used in structured programming classes.
  - CC (Calculus Calculator), by D. Meridith
  - Peanut Software – Plot
  - MBL tools
  - spreadsheets

18. *What was the effect of the use of technology?*

(a) *Interfered with group dynamics* (5 responses)

(b) *Improved group dynamics* (31 responses)

(c) *No effect* (4 responses)

(d) *Different effects on different groups* (2 responses)

  - TI-81 improved group dynamics, ISETL had no effect.
  - This semester ISETL is giving them something to unite and complain about throughout the campus.

- Focused the group dynamics, provided the need for and focus of group work done effectively.
- Its use is necessary in the courses I teach since they are computer programming or literacy classes.
- It is the tool which facilitates exploration — the boxes of discovery and generalization.
- Technology improved group dynamics in some but had no effect on others.

19. *How would you describe your cooperative learning experiences? Circle one.*

   (a) *Generally too problematic to continue cooperative learning* (no responses)

   (b) *Some problems, but some positive student gains* (16 responses)

   (c) *Generally successful* (25 responses)

   One wrote that it is too early to assess.

20. *How is the attrition rate affected by cooperative learning? Circle one.*

   (a) *Fewer students drop courses in which cooperative learning is a major feature.* (17 responses)

   (b) *More students drop such a course.* (2 responses)

   (c) *Cooperative learning seems to have no effect on attrition.* (15 responses)

   Some felt they did not have enough information to answer this.

21. *How did most students react to cooperative learning?*

   (a) *Very positively* (13 responses)

   (b) *Positively* (25 responses)

   (c) *Neutral* (4 responses)

   (d) *Negatively* (2 responses)

   Please send copies of any students' written comments that you are willing to share.

22. *What do you think that your students have gained from their cooperative learning experiences?*

- too early to tell

- learning to deal with others on many levels — getting in the habit of working on math with others

- a greater sense of self worth; appreciation for others and their ideas; a greater understanding of ideas/concepts in mathematics/statistics; how to get along with others; a more positive attitude about mathematics and learning mathematics; a more positive attitude about computers and their use

- All problems are not "easy." Working together is helpful. Answers should be explained. There is more than one correct approach to solving a problem (many more than incorrect). More students can be successful in math with this approach. They don't immediately fall behind (as they can on their own).

- broader, more in-depth understanding; learned social skills

- interaction

- confidence in working as a team

- improved understanding of mathematics, better students that help poorer students through discussion, teaching learning and defending positions. Experience working with others and solving problems that arise in the process.

- Sometimes this helped them develop some concepts better, also by exchange of ideas I would say the learning was more uniform and some possible questions that students might have had were answered.

- They learn to talk mathematics, to listen to others, to coordinate their needs with the needs of others, to be patient, to appreciate other's help, ...

- learn more, recognize *their* problems are general to the populations of students, communication is hard and one must work at it

- intellectual exchanges; time management; cooperation and collaboration; team spirit

- skills in working as part of a team; confidence working on longer, harder problems; verbal skills in "talking" mathematics

- learned to work with others; improved self-confidence; more ideas about specific problems; awareness of multiple ways of approaching a problem; more comfortable with technology — they had more of a sense of "belonging" to the department; more likely to approach me with a suggestion for changing something about the course; they can say "We discussed this in our group and we would like it if you would . . ."

- I hope the students will have gained a sense of responsibility for not only their learning, but also that of fellow students. The experience of explaining or developing concepts to those who don't have a good grasp of the concepts really firms up the concepts. We've all said "I really didn't understand it until I had to teach it."

- insight into problem-solving strategies and opportunities to develop tact and social skills

- experience working with other's personalities and ideas, further understanding of concepts from having explained their ideas to others, appreciation for the diverse skills people can bring to a project

- how to work in a team, assist one another, and be more responsible

- They put in more time on the course, and learn a team approach to academic work.

- Students learn better and get to know each other better. At the minimum, they are no worse off than without cooperative learning.

- very hard to sort that out from other aspects of the calculus course

- ability to learn from peers; social interactions; speed up class discussion of problems; get different points of view on some problems

- The mind set changes from *competition* (get ahead of my neighbor) to *cooperation* (together we can both come out better) — can only make a better world. They develop friends and meet more people; more LOVE, less HATE (competition).

- some exposure to working with others; practice in working with persons they may not care for; the realization that sometimes it helps to have exposure to other students' ideas

- the ability to work with other people whom you may not like; the nerve to assist yourself, to stand up for your rights

- I think some of them are finding out that they know more mathematics and so do their group members. They are finding out more about computers and how powerful computer software can be.

- a more positive attitude toward mathematics

- Classes have a mutually supportive atmosphere. Students talk about ideas; ideas have a concrete reality.

- learned concepts rather than formulas
- feeling that difficult problems can be solved
- Math can be a social activity.
- The students are learning to *think*, ask questions, and make connections.

23. *Please make any other comments that you think might be helpful.*

- A *major problem* students encounter is the ability to find a common meeting time — campus is mix of residents and commuters. Generally they think it takes more of their time to do the work.
- The class "lab" time is essential. They create and also solve problems there. A text with a good amount of *reading* material is necessary.
- Cooperative learning is a very positive experience; it is worth all the work.
- Good luck with this project. My experience suggests it is desperately needed because there are persons holding workshops, making presentations, passing as experts, etc., who have no experience/knowledge about use of groups at the college/university level. My experience is that all the *real* challenges (and maybe benefits) of using groups relate to their work *outside* the classroom.
- Most of my answers address grouping in my classes. I am not sure the *definition* of cooperative learning would always be a fit to my activities. I do firmly believe in "interaction" with content and peers as key to quality learning.
- I do like teaching more since I started finding ways to effectively use groups.
- I love cooperative learning and use it in several courses. More mature students seem to take to it more easily and use the opportunity more productively.
- At the beginning students have to be "sold" that cooperative learning will work for them.
- I just did an evaluation of the last 5 weeks of classes we have had and students' reactions were so positive that we took our desks out and now we have tables and chairs instead.
- I use teams in *all* my classes.
- You have not inquired about interactions with (or lack of) traditional courses and with colleagues not espousing these concepts and ideas. I find this to be a real problem.

- I needed a bigger class — three groups sometimes seemed like three students.

- allowed me to interact with students more; students feel more free to ask questions

- I am convinced that this method engages the students more than any I have tried.

Appendix C

Student Information Questionnaires

Some of the authors and some of the respondents to our survey have used the following forms (or variations of them) to collect information for group formation. These forms are completed by the students early in the semester. Some instructors have the students complete these questionnaires during the first class meeting, others prefer to wait until the end of the first week or the beginning of the second week so that the students have an opportunity to meet each other. Some instructors also have the students fill out time schedule sheets indicating class and work schedules.

C.1 Student Information Form for Calculus I

Calculus I Student Information Form

1. (PRINT) Name: _____
   Student ID#: _____

2. Major: _____

3. (PRINT) Address: _____
   Phone Number(s): _____

4. Age: _____

5. Gender (circle one):   male   female

6. What were your SAT (ACT) scores?   MA _____   V _____

7. What is your college GPA? _____   (Use high school GPA if no college GPA.)

8. What was the last math course you had? Where? When? Please give your grade.

    Course: _____   Where _____

    When: _____   Grade: _____

9. Have you had any calculus before? Circle one of the following:

    No   Yes, a few weeks   Yes, a semester or more

10. Have you worked in cooperative groups in any of your previous classes in high school or college? Circle one:

    No   Yes, high school   Yes, college   Yes, both

11. If you answered "yes" in item 10, were you given instruction on how to work cooperatively and develop social skills? Circle one:

                        No   Yes

12. Have you used a computer before? Circle one:

    No   Yes, a little   Yes, a lot

13. If you answered "yes" to item 12, what type of computer have you used before? Circle one:

    • IBM or compatible

    • Apple II

    • Macintosh

    • Other: _____

14. Have you written computer programs? Circle one:

    No   Yes, a little   Yes, a lot

15. Can you type? Circle one: No   Yes, a little   Yes, a lot

16. Please list names of anyone in this class you would like to work with.

17. Please list names of anyone in this class you would *not* like to work with.

C.2 Student Time Schedule Form

Calculus I Student Time Schedule

(PRINT) Name: _____

Please mark an X in any time slots when you have classes, work, or other obligations you *must* participate in, i.e., time you are *not available* to work with others outside of class or use the computer laboratory. Leave *blank* any hours you *could* choose (or negotiate) to use for work with your group or in the computer laboratory.

Hour	*Mon*	*Tue*	*Wed*	*Thu*	*Fri*	*Sat*	*Sun*
8 − 9							
9 − 10							
10 − 11							
11 − *Noon*							
Noon − 1							
1 − 2							
2 − 3							
3 − 4							
4 − 5							
5 − 6							
6 − 7							
7 − 8							
8 − 9							
9 − 10							
10 − 11							

C.3   Shorter Student Information Form

Calculus II Student Information Form

1. (PRINT) Name: _____

2. (PRINT Campus) Address: _____

3. (Campus) Phone Number(s): _____

4. (PRINT) Group Name in Calculus I? _____

5. What was your grade in Calculus I? _____

6. What is your grade point average? _____

7. List the names of any people you would particularly like to have in your group.

8. List the names of any people you would *not* particularly like in your group.

We make *no* promises about group composition.

C.4 An alternate student information form — Calculus I

Calculus I Student Information Form

1. (PRINT) Name: _____

2. Current address: _____

3. Current telephone number: _____

4. Gender: _____

5. Possible major(s): _____

6. Math SAT: _____   Verbal SAT: _____

7. What was the last mathematics course you took?

8. How much calculus have you studied? (Circle one of the following.)

   None   A few weeks only   One semester only   A full year

9. Have you used a computer?

   No   A few weeks only   A semester or more

10. Have you used an IBM-PC or compatible computer?

    No   A few weeks only   A semester or more

11. Can you type at all?

    No   A little   A lot

12. Advisor: _____

13. Which afternoons are you free after 1:30?

    Monday   Tuesday   Wednesday   Thursday   Friday

14. Is there anything else that you would like me to know?

C.5 An alternate student information form — Calculus II

Calculus II Student Information Form

1. (PRINT) Name: _____

2. Current address: _____

3. Current telephone number: _____

4. Gender: _____

5. Major or possible major(s): _____

6. Math SAT: _____   Verbal SAT: _____

7. What was your grade in Calculus I? _____

8. List the names of any students in the class that you would like in your group.

9. List the names of any students in the class that you would prefer not to have in your group.

10. Advisor: _____

11. Which afternoons are you free after 1:30?

    Monday   Tuesday   Wednesday   Thursday   Friday

12. Is there anything else that you would like me to know?

Appendix D

Class Participation Form

One of the authors has students fill out a Class Participation Form at the end of each class period. This form is returned to the students at the beginning of the next class period with comments and a score (based on ten points) for daily participation. At the end of the semester, the sum of each student's daily participation scores contributes 10 to 15% to the final course grade.

Class Participation Form

Name: .... **Course:** .... **Date:** ....

In what ways did you participate in this class?

... came to class prepared (+2)

... worked at computer (+1)

... answered questions (+2)

... participated in small group discussion (+3)

... offered my ideas (+3)

... listened to others (+1)

... asked questions (+2)

... attempted a problem (+2)

... worked at the board (+3)

Sum of positive points: ....

... came to class a bit late (-3)

... distracted others in my group (-3)

... came to class very late (-5)

... disrupted the class (-5)

Sum of negative points: ....

What did you learn by coming to this class?

What ideas or concepts are you confused about right now?

Appendix E

Information About Cooperative Groups on Course Syllabi

Some of the authors and respondents to our survey include expectations about working in cooperative groups on the course syllabus. These run the gamut from short simple statements of expectations to longer statements that suggest the scope of group activities and offer motivation for learning to work cooperatively. We include some samples here.

E.1 From a syllabus for Calculus I

Participation

The evaluation of participation includes consideration of regular class attendance, submission of journal entries each Friday, active engagement in class discussions and activities, and evidence of interest and involvement in all aspects of the course.

E.2 From a syllabus for Calculus II

E.2.1 Study Groups

You will do much of the work of this course in cooperative learning groups. It seems to work best if there are about three or four students in each group. You will be working with your group in the lab, on homework problems, on in-class activities, and even during quizzes and tests. Working well in a group is an important skill which is essential for many of the jobs for which our graduates apply. Some of you may enjoy the group work more than others. The objective of group work in this course is two-fold:

1. to give you moral support while you are working on problems or in the lab, and

2. to develop skills in working effectively as part of a team.

One of the primary objectives of this course is to help you to learn to think about and solve real-world problems using tools of mathematics. Working in your group, doing the lab activities, and talking about problems with your group members are all strategies to help you do this.

E.2.2   Assumptions about Homework

I suggest a lot of homework, and assume that:

- You are working on calculus problems regularly, almost daily. In particular, I expect that you will do some work outside of class *between every class.*

- A 4-credit class requires 8 to 12 hours of study time each week. It is most effective if your study time is distributed over the whole week, and not concentrated only on weekends.

- You are studying individually *and* working regularly with your group.

- You are responsible for your own learning, and so you will ask questions — in your group and in class discussion — about problems that have you stumped.

- You can't truly be stumped until you've given the problems a "royal try."

- I do not need to "collect and grade" homework in order to motivate you to do it. (If you need to turn in homework in order to be motivated, you may make a special arrangement with me.)

- You cannot completely understand a new concept until you have applied it to solving problems.

- Since symbolic computer systems (such as *Maple*) can do the "symbol crunching" for us, a college-level course in calculus requires that we give a lot of time and energy to problem situations that can be studied using the methods of calculus.

E.3 From a syllabus for Discrete Mathematics

Teaching approach

This course uses a hands-on approach instead of the traditional lecture approach. We integrate technology and cooperative learning into this Discrete Mathematics course.

The students will be placed into groups of about three to four students. Each group is to act as a team in which each person is accountable for each other person's learning with no one person dominating or doing all the work. Groups are to sit together for each class.

During the class periods conducted in the classroom the new concepts will be introduced in an intuitive way. The students will work together in groups on class tasks which are carefully designed to introduce the concepts. The students and the instructor will discuss the group's answers to the tasks and relevant questions. In the computer lab each group will share two computers and work on computer-based tasks. You will use the numeric, symbolic, and graphic capabilities of the computer software to draw conclusions and answer questions.

The students will use the computer algebra system *Derive* and the mathematical programming language *ISETL*. The idea of using computers is to help students to make mental constructions for the concepts and to free students from tedious calculations and the potential for arithmetic mistakes, so they can focus on interpretations and investigations.

This pedagogical approach

- promotes conceptual understanding of basic mathematical concepts,

- uses computers to help students discover mathematical ideas and learn mathematical concepts by constructing mental images on their own,

- uses cooperative learning where students work in small groups of three to four students,

- conveys the excitement and beauty of contemporary mathematics,

- captures students' imaginations and interest,

- uses an interactive approach, where students are active participants interacting with the computer, their groupmates, the workbook, and the instructor,

- makes students active learners by letting them explore mathematical ideas rather than simply applying the concepts for them.

Appendix F

Sample Grading Schemes

Students know those activities that are "really important" are included some-how in the calculation of course grades. Statements about course requirements and grading policies are usually included on course syllabi. We include several samples here from instructors at different institutions.

F.1 From syllabi for Calculus I & II

The following eight grades will be considered for each student:

- Group Grades:

    - Class participation: 5% (adjusted for individuals; see scale below)

    - Lab: 20% (adjusted for individuals; see scale below)

    - Homework: 15% (adjusted for individuals; see scale below)

    - Test 1: 15% (A: 90 – 100%; B: 80 – 89%; C: 70 – 79%; D: 60 – 69%; F: 0 – 59%)

    - Test 2 Group Average: 5% (A: 85 – 100%; B: 75 – 84%; C: 60 – 74%; D: 50 – 59%; F: 0 – 49%)

- Individual Grades:

    - Quizzes: 15% (A: 90 – 100%; B: 80 – 89%; C: 70 – 79%; D: 60 – 69%; F: 0 – 59%)

    - Test 2: 10% (A: 85 – 100%; B: 75 – 84%; C: 60 – 74%; D: 50 – 59%; F: 0 – 49%)

  − Final Exam: 15% (A: 85 − 100%; B: 75 − 84%; C: 60 − 74%;
  D: 50 − 59%; F: 0 − 49%)

Lab and Homework Scales:

  A: 90 − 100%; B: 80 − 89%; C: 65 − 79%; D: 55 − 64%; F: 0 − 54%

Class Participation Scale:

  A: 95 − 100%; B: 90 − 94%; C: 85 − 89%; D: 80 − 84%; F: 0 − 80%

Course grades will be determined as follows: Each letter grade will be converted
to a numerical value (A = 4.0, B = 3.0, . . . , F = 0.0), and the eight letter grades
will be averaged. For "borderline" grades, see "adjustments" below.

- A: average is in the closed interval [3.50, 4.00] (borderline if grade is "near"
  3.50)

- B: average is in the half-open interval [3.00, 3.50) (borderline if grade is
  near an end-point of this interval)

- C: average is in the half-open interval [2.00, 2.75) (borderline if grade is
  near an end-point of this interval)

- D: average is in the half-open interval [1.00, 2.00) (borderline if grade is
  near an end-point of this interval)

- F: average is less than 1.00, or if you have a failing grade average for your
  individual grades

Adjustments:

- A or B individual grade on Final Exam may raise final grade.

- *Only* low grade(s) is (are) a group grade(s) on Test 1 or Test 2 (or on both)
  may raise final grade.

- Score(s) within grade level "toward" the top may raise final grade. Score(s)
  within grade level "toward" the bottom may lower final grade.

- An adjusted Lab and/or Homework grade that is "low" may lower final
  grade.

F.2 From a syllabus for Calculus I

Grading Policy

Tests(3)	100 points each	30%
Quizzes (4)	25 points each	10%
Final examination	200 points	20%
Homework assignments	300 points	30%
Participation	100 points	10%

The first test will be a group test.

F.3 From a syllabus for Calculus I

Requirements

- **Regular attendance, participation, lab activities: 15%**

    You must *come to class regularly.* The material in this course has a well-deserved reputation for being difficult. If you miss a class, you are expected to find out what happened. The computer lab activities are designed to be an important way for you to learn the mathematical concepts. You are expected to work with the members in your group, and to seriously attempt the lab activities. *You will need to meet with your group outside of class.* Each week you will be asked to turn in evidence that you gave the lab activities a "royal try." Sometimes you will be asked to turn in individual work; other times I will ask you to give me your group response to some questions.

    At the end of each class period, I will ask you to fill out a *Class Participation / Self-Evaluation* form. I will use these forms to check attendance, respond to your self-evaluation, and give you a score for class participation.

    There will be occasional (unannounced) *quizzes* which you will be able to do in your small groups; I do not give "make-ups" for quizzes.

- **Tests: 50 – 60%**

    There will be *three tests.* The material to be covered on each test will be announced about two class periods prior to the scheduled test date. Tests are tentatively scheduled for the following dates:

    September 23 (Friday),

    October 17 (Monday),

    November 21 (Monday).

Ordinarily, I do not give make-up tests; exceptions to this policy will be considered on a case-by-case basis. At least one of these tests will have a group component; that is, you will do part of the test individually and part of it with your group.

Note: *These test dates are arranged so that there are two tests before mid-term (October 21). The last day to withdraw from a course is Friday, November 4. If you have concerns about your progress or ability to keep up with course assignments, please feel free to discuss these with me.*

- **Final Exam: 25 – 35%**

  The *cumulative final exam* is scheduled for Friday, December 16, from 1 to 3 pm. The final exam will have both individual and group parts.

F.4 From a syllabus for Discrete Mathematics

Grading

Your final grade will be based on the following items:

- **Group Grade**

  - **Group Assignments & Group Paper: 25%**

    * Group Assignments:
      Each week there will be a set of group assignments. Please hand in one copy of the group assignment in your group's designated folder. The signature of each group member who participated in solving the problems is to be recorded on the front sheet of the assignment. A group grade will then be assigned. A group should not allow a group member to sign if he or she did not participate in doing that homework.

    * Group Paper:
      Each group will write a paper on one topic in mathematics (e.g., paradoxes). I will provide you with the topic and guidelines.

  - **Test #1: 10%**
  - **Test #3: 10%**

- **Individual Grade**

  - **Test #2 + Test #4: 15%**
    If you miss a test (within the limits of allowable excused absences), you must contact the instructor to arrange for a make-up test within two days before or after the missed test; otherwise your grade for that test will be zero.

- **Individual Assignments & Paper: 15%**

  * Individual Assignments:
    Each week there will be a set of individual assignments. Please hand in your solutions in a folder. (No more than two assignments will be accepted late).

  * Individual Paper:
    I will provide you with a list of mathematicians, philosophers, and scientists. Each student will write a short biography of one of these persons and her or his contribution to mathematics. You will need to develop a bibliography, and are expected to use the library resources.

- **Comprehensive Final Examination: 20%**

  You may not miss the final exam. If you do so, you will fail the course.

- **Weekly Journal: 5%**

  Each student must keep a journal which is to be handed in each Friday. I will return the journals on Monday or Wednesday.

  I expect each journal to consist of two sections. The first section should contain a few sentences indicating how things are going, what's working and what's not working on the individual as well as on the group level. In the second section, write in your own words about the new concepts you have learned in the past week.

Bibliography

This bibliography offers a broad sampling of cooperative learning literature. The abstracts are selected from the ERIC database under different descriptors. In the parenthesis at the end of each entry is the ERIC number.

Alo, R. A. (1973). A collaborative learning approach for undergraduate numerical mathematics. *Paper presented at the Conference on Computers in the Undergraduate Curricula (4th, Claremont, California, June 18–20.* (ERIC Document Reproduction Service No. ED084157)

> Described is an undergraduate numerical analysis course organized around projects and tasks assigned to student teams. Most teams had five students within which the student with the most computer programming experience assumed the leadership role. The leaders' responsibility extended to distribution of work assignments and coordination of group interaction. Intragroup cooperation, leadership or lack of leadership and assignment of individual final grades are among the topics and problems discussed.

Alvarez, L., & Others. (1993). Calculus instruction at New Mexico State University through weekly themes and cooperative learning. *Primus, 3*(1), 83–98. (ERIC Document Reproduction Service No. EJ467790)

> Working in groups on weekly themes, students discover calculus concepts through assigned readings and written reports. Provides an outline of the course content for two courses and how the courses are organized. An appendix contains the complete text of two themes on curve sketching and applications of the definite integral.

Anderson, B. J. (1991). Community colleges: Promises or preclusions. *Paper presented at the Annual Convention of the American Mathematical Association of Two-Year Colleges.* 17th, Seattle, WA, November 7–10. (ERIC Document Reproduction Service No. ED351045)

Since nearly 10% of the students in the US. who receive doctorates in the mathematical sciences begin their undergraduate studies in two–year institutions, it is clear that these schools are a significant part of the mathematics educational pipeline. Yet, minority students enrolled in two-year colleges are one-fifth as likely to earn a bachelor's degree as those who start out in four-year schools. Given that approximately 50% of minorities in college are enrolled in two-year institutions, the role of these colleges in increasing minority participation in mathematics-related fields cannot be overstated. Mathematics teachers and professors, and those charged with facilitating the learning process, are the major change agents for improving the delivery of mathematics for minorities, as well as for all students. Proven change strategies include: (1) supporting a paradigmatic shift which asserts that all students can and must learn mathematics, and that minorities can succeed in mathematics-based fields; (2) setting up articulation and collaborative programs that make transfer from two- to four-year institutions smoother; (3) encouraging the best students to go into teaching; (4) intensifying minority recruitment; (5) promoting mathematics within minority communities by highlighting the successes of two-year college students in these communities; (6) restructuring remedial courses to incorporate cooperative learning, peer tutoring, and computer-assisted instruction; (7) setting numerical targets for minority student transfer; (8) establishing partnerships with industry; (9) seeking financial and human resources from government; and (10) promoting the teaching and learning function in mathematics.

Aronson, E., Blaney, N., Stephan, C., Sikes, J., & Snapp, m. (1978). *The jigsaw classroom.* Beverly Hills, CA: Sage.

Artzt, A. F. & Newman, C. M. (1990). Implementing the standards. Cooperative learning. *Mathematics Teacher, 83(6)*, 448–452. (ERIC Document Reproduction Service No. EJ415535)

Reviewed are the basic principles of cooperative learning including a rationale for its use and the formation of cooperative learning groups in the classroom. Examples of the application of this teaching method to mathematics teaching are discussed.

Artz, A. F., & Newman, C. M. (1990b). *How to use cooperative learning in the mathematics class.* Reston, Va.: The Council.

Balacheff, N. (1990). Towards a *problématique* for research on mathematics teaching. *Journal for Research in Mathematics Education, 21*(4), 258–272.

Berard, A. D., Jr. (1992). The use of small axiom systems to teach reasoning to first-year students. *Primus, 2(3)*, 265–277. (ERIC Document Reproduction Service No. EJ454997)

Uses small axiom systems to teach logic and reasoning to first year mathematics students. Incorporates a collaborative technique to train students to write formal arguments in a non intimidating atmosphere by applying logic informally to small axiomatic systems. Provides examples of student-designed systems.

Berg, K., F. (1993). Structured cooperative learning and achievement in a high school mathematics class. *Paper presented at the Annual Meeting of the American Educational Research Association*, Atlanta, GA. (ERIC Document Reproduction Service No. ED364408)

This study of college-bound 11th graders assessed the feasibility and effectiveness of instruction that used a structured cooperative learning technique. The students worked in dyads with scripts that contained two learning situations with two roles: (1) explainer and checker; and (2) solver and checker. Both students then worked on summary questions and homework. Verbal interaction influenced learning and appeared to be a mediator of the effects of student characteristics on achievement. Specifically, the study focused on two questions: (1) Can an effective program using dyadic studying techniques be designed for a high school course in higher mathematics; and (2) When high school students are trained to use a dyadic studying strategy for learning from their text, what is the nature of their verbal interaction and does this interaction change over time? Two groups were compared using the same texts, tests, and teacher. Both questions were answered affirmatively and supported statistically. The study concluded that: (1) students can be expected to respond positively to the experience and to work cooperatively and productively together; and (2) 94% of the time students had on-task interaction. Numerous tables contain specific statistical information. Contains 47 references.

Bossert, S. T. (1988–1989). Cooperative activities in the classroom. *Review of Research in Education, 15*, 225–250.

Bohlmeyer, E. M., & Burke, J. P. (1987). Selecting cooperative learning techniques: A constructive strategy guide. *School Psychology Review, 16(1)*, 36–49.

Brechting, S. M. C. & Hirsch, C. R. (1977). The effects of small group- discovery learning on student achievement and attitudes in calculus. *MATYC-Journal, 11(2),* 77–82.   (ERIC Document Reproduction Service No. EJ162919)

> Compared were the effects of two modes of instruction in the calculus: small group discovery and traditional lecture-discussion. The discovery mode was more effective in producing successful achievement in areas of manipulative skills, there were no differences in achievement as measured by a concepts test.

Cobb, P. & Steffe, L. P. (1993). The constructivist researcher as a teacher and model builder. *Journal for Research in Mathematics Education, 14*(2), 83–94.

Cohen, E. G. (1982, July). Sex as a status characteristic in a cooperative math-science curriculum. *Paper presented at the Conference of the International Association for the Study of Cooperation in Education,* Provo, Utah.

Cohen, E. G. (1986). *Designing group-work: Strategies for the heterogeneous classroom.* New York: Teachers College Press.

Cohen, E. G. (1994). Restructuring the Classroom: Conditions for Productive Small Groups. *Review of Educational Research,* 64(1), 1–35.

Cohen, E., Lotan, R., & Catanzarite, L. (1990). Treating status problems in the cooperative classroom, In S. Sharan (Ed.), *Cooperative learning: Theory and research,* 203–229. New York: Praeger.

Conciatore, J. (1990). From flunking to mastering calculus: Treisman's retention model proves to be "Too Good" on some campuses. *Black Issues in Higher Education, 6,* 5–6. (ERIC Document Reproduction Service No. EJ404651)

> This paper describes the development of a model for improving the calculus achievement of minority group college students currently used by 25 institutions of higher learning. Utilizes group study and an "honors class," rather than a remedial approach.

Cooper, L., Johnson, D. W., Johnson, R., & Wolderson, F. (1980). Effects of cooperative, competitive, and individualistic experiences on interpersonal attraction among heterogeneous peers. *Journal of Social Psychology, 111,* 243–252.

Davidson, N. (1980). Small-group learning and teaching in mathematics: An introduction for non-mathematicians. In S. Sharan, P. Hare, C. D. Webb, & R. Hertz-Lazarowitz (Eds.), *Cooperation in education.* Provo, Utah: Brigham Young University Press.

Davidson, N. (1985). Small-group learning and teaching in mathematics: A selective review of the research. In R. E. Slavin, S. Kagan, R. Hertz-Lazarowitz, C. Webb, & R. Schmuck. *Learning to cooperate, cooperate to learn,* 221–230. New York: Plenum.

Davidson, N. (1989a). Cooperative learning in mathematics. *Cooperative Learning, 10*, 2–3. (ERIC Document Reproduction Service No. EJ420759)

> Small group cooperative learning can be applied with all age levels of mathematics students. A community of learners actively working together enhances each person's mathematical knowledge, proficiency, and enjoyment. Classroom procedures for cooperative mathematics lessons are outlined.

Davidson, N. (1989b). Cooperative learning and mathematics achievement: A research review. *Cooperative Learning, 10*, 15–16. (ERIC Document Reproduction Service No. EJ420760)

> The current status of research and development on cooperative learning in mathematics is discussed in an interview with Neil Davidson. The focus is on significant achievement differences favoring small group procedures. Key issues discussed here are cooperative learning and problem-solving skills, and students' need for some kind of reward.

Davidson, N. (1990a). The small-group discovery method in secondary and college level mathematics. In N. Davidson *Cooperative learning in mathematics: A handbook for teachers.* Menlo Park, California: Addison-Wesley Publishing Co.

Davidson, N. (1990b). *Cooperative Learning in Mathematics: A Handbook for Teachers.* Addison-Wesley Publishing Company, Inc., Addison-Wesley Innovative Division, 2725 Sand Hill Rd., Menlo Park, CA 94025. (ERIC Document Reproduction Service No. ED335227)

> Small group cooperative learning provides an alternative to both traditional whole-class expository instruction and individual instruction systems. The procedures described in this volume are realistic, practical strategies for using small groups in mathematics teaching and learning with methods applicable to all age levels, curriculum levels,

and mathematical topic areas. Included are: (1) an introduction and overview to orient the reader; (2) problem solving and exploration with manipulative materials in groups of four; (3) a bilingual integrated mathematics/science program addressing classroom status; (4) team learning approaches based upon individual accountability and team recognition; (5) a general conceptual model of cooperative learning with a detailed discussion of its basic components; (6) three computer-based cooperative learning strategies for classroom use; (7) various learning activities for the initiation of a cooperative classroom setting; (8) procedures for group problem solving and inquiry in algebra, geometry, and trigonometry; (9) a narrative on cooperation in heterogeneous group mathematics in the Netherlands; (10) group interactions in algebra and calculus using computers for problem solving; (11) free exploration and guided discovery in cooperative groups with examples from calculus; (13) issues affecting the use of cooperative learning in mathematics with emphasis on teachers' decision making and factors affecting implementation; and (14) appendices that include information about the sponsoring organizations, annotated survey responses from classroom teachers, and annotated listing of 38 resource materials with contact person.

Davidson, N. & O'Leary, P. W. (1990). How cooperative learning can enhance mastery teaching. *Educational Leadership, 47(5)*, 30–33. (ERIC Document Reproduction Service No. EA524132)

> Transforms the debate over cooperative learning and Hunter's mastery teaching model by illustrating how both approaches reinforce each other. Mastery teaching synthesizes the most rewarding aspects of traditional expository instruction, while cooperative learning breathes life into that teaching by inviting both students and teachers to become idea "coproducers." Includes 17 references.

Davidson, N., & others. (1990). Staff development for cooperative learning in mathematics. *Journal of Staff Development, 11*, 12–17. (ERIC Document Reproduction Service No. EJ430612)

> The article discusses four models of staff development for cooperative learning in mathematics: graduate seminar with weekly meetings; workshop enhanced by support systems and follow-up; school-based staff development with monthly meetings; and projects involving collaboration between university and school district personnel. The article lists 10 important conclusions about staff development.

Davidson, N. & Kroll, D. L. (1991). An overview of research on cooperative learning related to mathematics. *Journal for Research in Mathematics*

Education, 22, 362–365. (ERIC Document Reproduction Service No. EJ 436598)

> Increased use of cooperative learning methods is a visible change in mathematics education in the last decade. Research questions on cooperative learning concerning different models employed, their effectiveness compared to traditional methods of instruction, their effects on student achievement, and cognitive and affective benefits gained during student learning are reviewed.

Davy, B. (1983). Think aloud — modeling the cognitive process of reading comprehension. *Journal of Reading, 27*(1), 44–47.

Dees, R. (1989). Cooperative mathematics lesson plans. *Cooperative Learning, 10*, 32–40. (ERIC Document Reproduction Service No. EJ429362)

> The article presents several classroom-tested model cooperative mathematics activities, ranging from simple to highly structured, for all grade levels. Some of the activities are creating and solving problems from the newspaper; playing team-building games; exploring with fractions; and working on grouping and informal computation using a jar full of candy kisses.

Dees, R. (1991). The role of cooperative learning in increasing problem-solving ability in a college remedial course. *Journal of Research in Mathematics Education, 22(5)*, 409–421. (ERIC Document Reproduction Service No. EJ436601)

> Students ($n = 105$) enrolled in a college remedial mathematics course participated in a one-semester experiment to determine whether cooperative learning would help students increase their problem-solving skills in mathematics. Results indicated significant differences in favor of students using cooperative learning in solving word problems in algebra and geometry.

Delgado, M. T. (December 1987). *The effect of a cooperative learning strategy on the academic behavior of Mexican-American children.* Doctoral diss. Stanford University.

De Vries, D., & Edwards, K. (1974). *Cooperation in the classroom: Towards a theory of alternative reward-task classroom structures.* Paper presented at the annual meeting of the American Educational Research Association, Chicago, IL.

Dishon, D., & O'Leary, P. (1984). *A guidebook for cooperative learning: A technique for creating more effective schools.* Holmes Beach, FL: Learning Publications.

Dubinsky, E. (1989a). Constructive aspects of reflective abstraction in advanced mathematical thinking. In L. P. Steffe (Ed.), *Epistemological Foundations of Mathematical Experience.* New York: Springer-Verlag.

Dubinsky, E. (1989b). A theory and practice of learning college mathematics. In A. Schoenfeld (Ed.) *Advanced Mathematical Thinking.*

Dubinsky, E. & Schwingendorf, K. (1990). Constructing calculus concepts: Cooperation in a computer laboratory. In Carl Leinbach et al (Eds.) *Laboratory Approach to Teaching Calculus.* Mathematical Association of America Notes Series, *20,* Mathematical Association of America, 47–70.

Emley, W. J. (July 1987). *The effectiveness of cooperative learning versus individualized instruction in a college level remedial mathematics course, with relation to attitudes toward mathematics and Myers-Briggs personality type.* Doctoral diss. University of Maryland College Park.

Finkel, D. L. & G. S. Monk (1983). Teachers and learning groups: Dissolution of the Atlas Complex, In C. Bouton & R. Y. Garth (Eds.) *Learning in groups.* New Directions for Teaching and Learning, No. 14. San Francisco: Jossey-Bass, 83–96.

Foyle, H. C., & Lyman, L. (1989). Cooperative learning: research and practice. *Paper presented at the Rocky Mountain Regional Conference for the Social Studies, Phoenix, Arizona.* (ERIC Document Reproduction Service No. ED308131)

> Jigsaw, a form of cooperative learning, was researched by Aronson (1978). Later, Slavin (1981) adapted Jigsaw to Student Team Learning and called it Jigsaw II. Jigsaw currently shows the least achievement gains among the various cooperative methodologies. Nonetheless, it is a viable methodology and is useful for covering and reviewing material. This document provides a way of implementing a Jigsaw-type lesson about cooperative learning. Student instructions, questions for each group to answer, and 16 abstracts selected from the ERIC database under the descriptor of "Cooperative Learning" are included. The abstracts are to be used by student groups as evidence or research findings from which they make their decisions.

Garofalo, J. (1987). Metacognition and School mathematics. *Arithmetic Teacher*, 9(May), 22–23.

Garofalo, J. (1989). Beliefs and their influence on mathematical performance. *Mathematics Teacher.* (October), 502–505.

Garofalo, J. & Lester, F. K. (1985). Metacognition, cognitive monitoring, and mathematical performances. *Journal for Research in Mathematics Education,* 16, 163–176.

Gonzalez, A. (1980). *Classroom cooperation and ethnic balance.* Unpublished doctoral dissertation, University of California, Santa Cruz.

Good, T. L., Grouws, D., & Ebmeier, H. (1983). *Active mathematics teaching.* New York. Longman.

Good, T. L., & Marshall, S. (1984). Do students learn more in heterogeneous or homogeneous groups? In P. E. Peterson, L. C. Wilkinson, & M. T. Hallinan (Eds.), *Student diversity and the organization, process, and use of instructional groups in the classroom.* New York: Academic Press.

Graves, T. (1990). Cooperative learning and academic achievement: A tribute to David and Roger Johnson, Robert Slavin, and Shlomo Sharan. *Cooperative Learning, 10,* 13–16. (ERIC Document Reproduction Service No. EJ420767)

> An overview is presented of the work and theories of three teams of researchers who have focused on cooperative learning. Cooperative learning has become the outstanding example of an educational innovation in which practice is informed by research; the collective leadership of these teams has advanced its use.

Graves, T. (1991). The controversy over group rewards in cooperative classrooms. *Educational Leadership, 48,* 77–79. (ERIC Document Reproduction Service No. EJ424424)

> Suggests ways to minimize the negative effects of extrinsic group rewards in cooperative classrooms, explains how to use intrinsic rewards, and outlines conditions calling for extrinsic rewards. The "social rewards" of working cooperatively probably enhance intrinsic motivation and are among the great advantages of employing cooperative learning strategies.

Graves, N., & Graves, T. (1990). *What is cooperative learning? Tips for teachers and trainers,* (2nd ed.). Santa Cruz: Cooperative College of California.

Grossman, A. S. (January 1985). Mastery learning and peer tutoring in a special program. *Mathematics Teacher, 78,* 24–27.

Gura, K. (1992). Liberal arts mathematics: Probability and calculus. *Primus, 2(2)*, 155–164. (ERIC Document Reproduction Service No. EJ453541)

> Presents one model for a liberal arts mathematics course that combines probability and calculus. Describes activities utilized in the course to heighten students' interest and encourage student involvement. Activities include use of visualization, take-home tests, group problem solving, research papers, and computer usage with DERIVE computer software.

Hertz-Lazarowitz, R. (1990, April). *An integrative model of the classroom: The enhancement of cooperation in learning.* Paper presented at the Annual Meeting of the American Educational Research Association, Boston, MA. (ERIC Document Reproduction Service No. ED322121)

> Cooperative learning aims to enhance students' on task interactive behaviors in the classroom. Observation in Israeli elementary schools has indicated that interactive behavior of students in their learning sequence holds potential for quality cooperation and help among children, but that teachers lack the skills to structure learning tasks that will enhance a high level of cooperation. Based on prior research an integrative model of the classroom was developed, and detailed developmental stage taxonomy was suggested to describe, explain, guide, and predict students' cooperative interaction in all types of classroom structures. The model is based on six related dimensions: (1) the physical organization of the classroom; (2) the nature of the learning task; (3) the instructional mode of the teacher; (4) the communication pattern of the teacher; (5) student social-learning behaviors; and (6) students' academic-cognitive behavior. This model and the research that followed indicate that variation in learning task structure and teacher behavior are the main factors in shaping students' behaviors. The inclusion of cooperative learning as part of the daily experience in the classroom is proposed.

Hertz-Lazarowitz, R., & Davidson, J. (1990). *Six mirrors of the classroom: Pathway to a cooperative classroom.* Westlake Village, CA: Rajo Press.

Hom, H. L., Berger, Duncan, M., Miller, A., & Blevin, A. (1990). *The influence of cooperative reward structures on intrinsic motivation.* Springfield, MO.: Southwestern Missouri State University.

Hoyles, C. (1985). What is the point of group discussion in mathematics? *Educational Studies in Mathematics, 16*, 205–214.

Johnson, D. W. (1982). *Experiments to Attain Full Participation of Handicapped Students in the Regular Classroom. Final Report.* Minnesota Univ., Minneapolis. (ERIC Document Reproduction Service No. ED245514)

Reprints of 17 studies on approaches to ensuring full participation of handicapped students in the regular classroom are presented. The studies, carried out over a 3-year period, were intended to examine evidence on the efficacy of mainstreaming, with particular emphasis on the role of competitive, and individualistic learning experiences on friendship, interpersonal attraction, performance, achievement, and relationships between handicapped and nonhandicapped students. The studies point out the value of the cooperative learning approach in securing active participation of handicapped students, generalizing positive relationships to free time situations, promoting achievement and self-esteem of handicapped students, and promoting the ability of nonhandicapped students to take the perspective of their handicapped peers. The cooperative approach is said to be easily developed and implemented and to result in benefits for both handicapped and nonhandicapped students.

Johnson, D. W., & Johnson, R. T. (1980). *Promoting Constructive Student-Student Relationships through Cooperative Learning.* Minnesota Univ., Minneapolis. National Support System Project. (ERIC Document Reproduction Service No. ED249216)

This module (part of a series of 24 modules) is on the impact of interaction among students in a learning situation on achievement, cognitive development, and social development. The genesis of these materials is in the 10 "clusters of capabilities," outlined in the paper, "A Common Body of Practice for Teachers: The Challenge of Public Law 94-142 to Teacher Education." These clusters form the proposed core of professional knowledge needed by teachers in the future. The module is to be used by teacher educators to reexamine and enhance their current practice in preparing classroom teachers to work competently and comfortably with children who have a wide range of individual needs. The module includes objectives, scales for assessing the degree to which the identified knowledge and practices are prevalent in an existing teacher education program, and self-assessment test items. Articles are appended on influences of peer interaction and school outcomes, the social integration of handicapped students, and cooperative instructional games.

Johnson, D. W., & Johnson, R. T. (1984). Cooperative small-group learning. *Curriculum Report, 14* (ERIC Document Reproduction Service No. ED249625)

In cooperative learning, as opposed to competitive and individualistic learning, students work together to accomplish shared goals. It is the most important of the three types of learning, but least used. Research indicates students will learn more, like school better, like each other better, and learn more effective social skills when cooperative learning is used. It is not simply a matter of putting students into groups to learn, but involves positive interdependence, face-to-face interaction, individual accountability, and appropriate use of interpersonal and small group skills. Among strategies necessary to implement cooperative learning are: clearly specifying lesson objectives; making plans about the cooperative learning group; explaining the academic task and cooperative goal structure to the students; monitoring effectiveness and providing assistance with interpersonal and group skills; and evaluating student achievement and helping them assess how well they collaborated with each other. The principal's role includes structuring and managing a support system for teachers, with teachers providing the basic support for each other. The report describes procedures that sample school districts followed in implementing cooperative learning, and gives the addresses of this Cooperative Learning Center, a resource for information.

Johnson, R. T., & Johnson, D. W. (1985). Student-student interaction: Ignored but powerful. *Journal of Teacher Education, 36*, 22–26. (ERIC Document Reproduction Service No. EJ322872)

Although cooperative learning evidences the greatest potential for engendering student learning, it is used no more than 20 percent of the time in most classrooms. This article reviews research on cooperative, competitive, and individualistic structures used by teachers and discusses the implications for changing teacher preparation practices.

Johnson, R. T., & Johnson, D. W. (1987a). How can we put cooperative learning into process? *Science Teacher, 54*, 46–48. (ERIC Document Reproduction Service No. EJ358518)

Discusses some of the benefits of having students work cooperatively toward a shared goal. Suggests that teachers become involved in research in this area and become associated with the National Science Teachers Association's Every Teacher a Researcher (ETR) program. Lists various types of research projects teachers could undertake.

Johnson, D, W., & Johnson, R. T. (1987b). *Learning together and alone,* (2nd ed.) Englewood Cliffs, N.J.: Prentice-Hall.

Johnson, D. W. & Johnson, R. T. (1987c). *Learning Together and Alone: Cooperation, Competition and Individualization.* Englewood Cliffs, NJ: Prentice Hall Inc.

Johnson, D. W., & Johnson, R. T. (1989). *Cooperation and competition: Theory and research.* Edina, Minn.: Interaction Book Co.

Johnson, D. W., & Johnson, R. T. (1990). Social skills for successful group work. *Educational Leadership, 47,* 29–33. (ERIC Document Reproduction Service No. EJ400495)

> People do not know instinctively how to interact effectively with others. For cooperation to succeed, students must get to know and trust one another, communicate accurately and unambiguously, accept and support one another, and resolve conflicts constructively. A seven-step recommended procedure is outlined.

Johnson, D. W., & others. (1984). *Circles of Learning: Cooperation in the Classroom.* Association for Supervision and Curriculum Development, Alexandria, Va. (ERIC Document Reproduction Service No. ED241516)

> Cooperative learning processes have been rediscovered and are being used throughout the country on every level. The basic elements of cooperative goal structure are positive interdependence, individual accountability, face-to-face interaction, and cooperative skills. The teacher's role in structuring cooperative learning situations involves clearly specifying lesson objectives, placing students in productive learning groups and providing appropriate materials, clearly explaining the cooperative goal structure, monitoring students, and evaluating performance. For cooperative learning groups to be productive, students must be able to engage in the needed collaborative skills. Cooperative skills and academic skills can be taught simultaneously. The implementation of collaborative professional support groups among educators. Both the success of implementation efforts and the quality of life within most schools depend on teachers and other staff members cooperating with each other. Support for the program takes as careful structuring and monitoring as does cooperative learning.

Johnson, D. W., Johnson, R., & Holubec, E. (1987). *Structuring cooperative learning: Lesson plans for teachers.* Edina, Minn.: Interaction Book Company.

Johnson, D. W., Johnson, R., & Holubec, E. (1988). *Cooperation in the classroom.* Edina, Minn.: Interaction Book Company.

Johnson, D.W., Johnson, R., & Maruyama, G. (1983). Interdependence and interpersonal attraction among heterogeneous and homogeneous individuals: A theoretical formulation and a meta-analysis of the research. *Review of Educational Research, 52,* 5–54.

Johnson, D. W., Maruyama, R., Johnson, R., Nelson, D., & Skon, L. (1981). Effects of cooperative, competitive and individualistic goal structure on achievement: A meta-analysis. *Psychological Bulletin, 89,* 47–62.

Johnson, H. A. (1985). *The effects of the groups of four cooperative learning model on student problem-solving achievement in mathematics.* Doctoral dissertation, University of Houston.

Kagan, S. (1985). *Cooperative learning.* Mission-Viejo, CA: Resources for Teachers.

Kagan, S. (1989a). The structural approach to cooperative learning. *Educational Leadership, 47,* 12–15.

Kagan, S . (1989b). *Cooperative learning resources for teachers.* San Juan Capistarano, California: Resources for Teachers.

Kagan, S. (1992). *Cooperative Learning.* Mission Viejo, CA: Resources for Teachers.

Kaufer, L. (1976). Personalizing large group instruction. *MATYC-Journal, 10(3),* 156–157. (ERIC Document Reproduction Service No. EJ152012)

> The organization of a mathematics class involving large lecture instruction along with small group work and remedial work is described.

Kraft, R. G. (1985). Group-inquiry turns passive students active, *College Teaching,* 33(4), 149–154.

Krantz, G. S. (1993). How to teach mathematics: A personal perspective. *American Mathematical Society,* Providence, RI.

LeGere, A. (1991). Collaboration and writing in the mathematics classroom. *Mathematics Teacher, 84(3),* 166–171. (ERIC Document Reproduction Service No. SE547339)

> Described are classroom strategies chosen to elicit greater involvement by students in the learning process and to furnish opportunities for practice in critical thinking. The advantages and disadvantages of this teaching approach are discussed.

Levy, B. N. (1990). A MathCAD exploration: Hunting for hidden roots. *Mathematics Teacher, 83(9)*, 704–708.

> Discussed is the use of a program called MathCAD which allows students to solve higher order polynomial equations and geometry problems. The use of cooperative solving is emphasized. Included are graphs and part of a printout generated while solving problems with MathCAD.

MacLeod, S. (1992). Ideas in practice: Writing the book on fractions. *Journal of Developmental Education, 16(2)*, 26–28. (ERIC Document Reproduction Service No. EJ454852)

> Describes a project in which students in a community college basic mathematics class worked collaboratively to write their own book on fractions. Students reinforced their math and cooperative skills, gained confidence, organized and revised their ideas, and created their own examples. Relates teacher and student responses to the project.

Madden, N. A., & Slavin, R. E. (1983). Cooperative learning and social acceptance of mainstreamed academically handicapped students. *Journal of Special Education, 17*, 171–182.

Male, M., Johnson, R., Johnson, D., & Anderson, M. (1986). *Cooperative learning and computers: An activity guide for teachers.* Santa Cruz, CA: Educational Apple-cations.

Marshall, G. (1990). A changing world requires changes in math instruction. *Executive Educator, 12*, 23–24. (ERIC Document Reproduction Service No. EJ411655)

> In response to our technological society, the National Council of Teachers of Mathematics recommends decreasing drill and reliance on multiple-choice tests and increasing problem solving and real-world applications. Other recommendations include simulation software to provide mathematical challenges, collaborative learning, and alternative assessment methods.

Maskit, D., & Hertz-Lazarowitz, R. (1986, April). *Adults in cooperative learning: Effects of group size and group gender composition on group learning behaviors.* Paper presented at the Annual Meeting of the American Educational Research Association, CA. (ERIC Document Reproduction Service No. ED279788)

The study described in this report investigated the effects of two context variables on small group learning-namely, group size and group gender composition-within an adult learning framework. In the study, the "revolving circle" method was innovated. In this design, discussions are around his/her task completion. The method was different classrooms in Haifa, Israel, taught by four teachers in literature and language arts. Students in each class were randomly assigned to groups of different size (three, four, or five group members) and different gender composition (majority male or female). Two trained female observers watched each classroom for six full periods of 90 minutes each, and coded behavior for 5 minutes using a checklist. Observed behaviors were grouped in six categories: listening and social interaction group maintenance, interactive summary, giving and requesting information, cooperative learning behaviors, and general learning behaviors. The study found that most of the significant differences occurred in odd-number groups. Groups of three members elicited more integrative summary and general learning behaviors, while groups of five members elicited more cooperative learning behaviors, listening, and social interaction. The study also found that cooperative learning behavior was significantly higher in groups with either gender majority, while giving information was highest in equal gender composition groups. The results of the study can be used in further research on group learning behavior.

Mathews, S. M. (1991). Group problem solving in the college mathematics classroom. *Primus, 6(4)*, 430–442. (ERIC Document Reproduction Service No. SE548658)

Describes the mechanics of group work in the college mathematics classroom specifically group formation, preliminary class work, class and group discourse, individual and group assignments, and impact on test taking. Includes examples from a first-semester calculus course.

Mavarech, Z. R. (July/August 1985). The effects of cooperative mastery learning strategies on mathematical achievement. *Journal of Educational Research, 78*, 372–377.

Mavarech, Z. R. (March/April 1991) Learning mathematics in different mastery environments. *Journal of Educational Research, 84*, 225–231.

Mayers, C. (1986). Teaching students to think critically. *Jossey-Bass Higher Education Series*. San Francisco: Jossey-Bass.

Moskowitz, J. M., Malvin, J. H., Schaeffer, G. A., & Schaps, E. (1983). Evaluation of a cooperative learning strategy. *American Educational Research Journal, 20*, 687–696.

Mumme, J., & Weissglass, J. (1989). The role of the teacher in implementing the "Standards." *Mathematics Teacher, 82*, 522–526. (ERIC Document Reproduction Service No. EJ406006)

> Answers several questions asked by teachers about the implementation of the "Curriculum and Evaluation for School Mathematics," including teacher role; textbook change; starting procedures; finding support groups; dealing with feelings; classroom change; and curriculum change.

Newmann, F. M., & Thompson. (1987). *Effects of cooperative learning on achievement in secondary schools: A summary of research.* Madison, WI.: University of Wisconsin, National Center on Effective Secondary Schools.

Noddings, N. (1985). Small groups as a setting for research on mathematical problem solving. In E. A. Silver (Ed.), *Teaching and learning mathematical problem solving*, 345–359. Hillsdale, NJ: Erlbaum.

O'Brien, M., & Chalif, D. (1991). *CHEMATH: A Learning Community in Science and Math.* Edmonds Community Coll., Lynnwood, WA. (ERIC Document Reproduction Service No. ED336157)

> The CHEMATH course at Edmonds Community College integrates instruction in algebra and chemistry. The course emphasizes problem-solving techniques and relies on active learning through the use of structured group process techniques. CHEMATH is team taught by science and math faculty and is intended primarily for students planning to major in science, math, or engineering, but who have a poor background in math and science. Students completing CHEMATH 105/110 receive the credit equivalent of both Chemistry 105 and Mathematics 110. The focus on small-group problem-solving skills allows students to learn to use each other as resources; to communicate and describe problems; to learn to take responsibility for their own learning; and to build support systems. This report consists primarily of presentation overheads describing the program, and includes sample comments by participating faculty and students; course syllabi for CHEMATH 105/110, Chemistry 105, and Mathematics 110; and sample quiz, group word problem set, lab exercise, and homework problem set.

O'Connor, J. E. (1993). *Evaluating the effects of collaborative efforts to improve mathematics and science curricula.* Paper presented at the Annual Meeting of the American Educational Research Association (Atlanta, GA, April 12–16). (ERIC Document Reproduction Service No. ED357083)

The Mathematics and Science Partnership Project (MSPP) is a 3-year collaborative effort among the U.S. Department of Education, IBM California Educational Partnership Program, California State University in Bakersfield, and the Kern High School District in Bakersfield (California) to impact mathematics and science curricula through the implementation of innovative technology. Thirteen teachers at six high schools have been provided with state-of-the-art computer networks and multimedia workstations, as well as mathematics and science software and training in implementing this technology. Some results of this effort are presented, following the first 2.5 years. Various evaluation methods have been used, including teacher journals, classroom observations, videotaping, questionnaires, student interviews, and a quasi-experimental comparison (not yet complete) of students using the technology with a group without access to the technology. The impact on teaching and learning has been documented. Teachers are using a more student-centered approach, and using more cooperative learning groups. Student motivation is improved, and students are enthusiastic about the technology. Implications of this collaborative effort are discussed. One table presents results of student surveys, and student interview questions are included.

O'Malley, C. E. & Scanlon, E. (1990). Computer-supported collaborative learning: Problem solving and distance education. *Computers and education, 15*, 127–136. (ERIC Document Reproduction Service No. EJ407233)

Discusses cooperative problem solving among Open University physics and math students and describes three studies that investigated how to design effective computer-based support for collaborative learning in distance education. The value of peer interaction for learning and problem solving is discussed, and students' attitudes to group work are examined. (28 references).

Owens, J. E. (1992). *Cooperative learning in secondary education: Research and theory.* Manuscript submitted for publication.

Page, W. (1979a). Questions in the round: An effective barometer of understanding. *Two-Year College Mathematics Journal, 10*, 35.

Page, W. (1979b). A small group strategy for enhanced learning. *American Mathematical Monthly, 86(10)*, 856–858. (ERIC Document Reproduction Service No. EJ215062)

A summary-review strategy for use in mathematics classes is presented. Specific objectives attainable through use of this strategy are outlined and two procedures for using the small-group strategy are given.

Page, W. (1984, July). *Knowledge transmission and acquisition: Cognitive and affective considerations.* Paper presented at the Sloan Foundation Conference on New Directions in Two Year College Mathematics, Atherton, CA. (ERIC Document Reproduction Service No. ED247996)

> Arguing that college mathematics education must be made more effective, especially for technology, engineering, mathematical sciences, and physical sciences students, this paper presents nine general principles to enhance math instruction for all students. Introductory material argues that changes in perception, attitudes, and role models are needed to realize the goals of integrating knowledge acquisition and knowledge utilization and exploring metacognitive instructional considerations. Next, a historical and futuristic overview is provided of important mathematical issues of the 20th century. Then, the general principles for mathematics instruction are presented, discussed, and illustrated with examples: (1) "what" one communicates in mathematics instruction includes the intrinsic nature and value of the discipline; (2) "how" one communicates goes beyond the exchange of ideas and information to long-lasting psycho social values; (3) math teachers must appreciate individual differences and their impact on learning styles; (4) multimodel representation of concepts has the potential for synergistic learning; (5) math principles should be presented as the basis for solving classes of problems; (6) students need to learn to reformulate and restructure problem representations; (7) teachers must anticipate and preempt students' misinterpretations; (8) control of knowledge must be appreciated as part of knowledge acquisition and accumulation; and (9) students must be responsible partners in an interactive and collaborative learning environment.

Palmer, J., & Johnson, J. T. (1989). Jigsaw in a college classroom: Effect on student achievement and impact on student evaluations of teacher performance. *Journal of Social Studies Research, 13,* 34–37. (ERIC Document Reproduction Service No. EJ406196)

> Examines a cooperative learning technique called Jigsaw that was used to determine whether college-level students taught by this method scored higher on a posttest than students who were not. Results showed no significant difference between those taught by the Jigsaw technique and those who were not.

Parker, R. (1984). Small group cooperative learning in the classroom. *OSSC Bulletin, 27,* 1984. (ERIC Document Reproduction Service No. ED242065)

> In contrast to the recent back-to-basics movement, which emphasizes rote-learning and the acquisition of mechanistic skills, small-

group cooperative learning emphasizes the development of thinking and problem-solving skills. It also seeks to minimize student anxiety and competition by creating an environment in which students feel safe to make and learn from mistakes. Research on cooperative learning suggests that the approach has proven effective in achieving both social and academic goals. As is clear from teachers' experiences with a technique in which students work together in "groups of four" randomly selected every two weeks, the approach requires a different role for teachers and students. Teachers must give up their dominant role in relation to their pupils' thinking, and students must learn to accept more responsibility for themselves and their peers. The approach also involves spatial reorganization, a heightened noise level, and ongoing student evaluation. Although experienced teachers who have used cooperative learning agree as to its advantages, successful implementation requires long-term commitment, support, and understanding from principals and colleagues. Two appendixes describe the most widely used cooperative learning models and discuss such models under two major headings: peer tutoring methods and group investigation methods.

Peterson, P. L., Wilkinson, L. C. & Hallinan, M. (1984). *The Social Context of Instruction. Group Organization and Group Processes.* Academic Press, Inc., Orlando, Florida. (ERIC Document Reproduction Service No. ED268075)

This book is an outgrowth of a conference funded by the National Institute of Education and held at the Wisconsin Center of Education Research in May 1982. A major theme of this volume of collected papers is how and in what ways grouping of students can be used effectively. Papers included are: (1) "Instructional groups in the classroom: Organization and processes" (P. L. Peterson and L. C. Wilkinson); (2) "Do students learn more in heterogeneous or homogeneous groups?" (T. L. Good and S. Marshall); (3) "Grouping and instructional organization" (S. T. Bossert, B. G. Barnett, and N. N. Filby); (4) "The social organization of instructional grouping" (J. E. Rosenbaum); (5) "First grade reading groups: Their formation and change" (R. Dreeben); (6) "Effects of race on assignment to ability groups" (A. B. Sorensen and M. Hallinan); (7) "Frameworks for studying instructional processes in peer-work-groups" (S. S. Stodolsky); (8) "Merging the process-product and the sociolinguistic paradigms: Research on small group processes" (P. L. Peterson, L. C. Wilkinson, F. Spinelli, and S. R. Swing); (9) "Student interaction and learning in small group and whole class settings" (N. M. Webb and C. M. Kenderski); (10) "Talking and working together: Status, interaction, and learning" (E. G. Cohen); (11) "The development of attention norms in ability groups" (D. Edgar and D. Felmlee); and (12) "Vygotskian perspectives on discussion processes in small group reading lessons" (K. Hu-pei Au and A. J. Kawakami). M. Hallinan presents a summary and implications.

Phoenix, C. Y. (1991). A four-strategy approach used to teach remedial mathematics in a freshman year program. *Community-Review, 7*, 45–52. (ERIC Document Reproduction Service No. EJ447621)

Describes the use of student verbalization and immediate feedback, cooperative learning, a concept/discovery-based approach, and creative classroom activities in a remedial mathematics class for first-year students at Medgar Evers College. Compares student achievement with that in other sections of the same course. Includes sample problems.

Piaget, J. (1926). *The language and thought of the child.* (M. Gabain, Trans.). London: Routledge and Kegn Paul, Ltd.

Piaget, J. (1932). *The moral judgment of the child.* (M. Gabain, Trans.). New York: Harcourt, Brace and World, Inc.

Piaget, J. (1950). *The psychology of intelligence.* (M. Piercy & D. E. Berlyne, Trans.), Routledge & Kegan Paul LTD, Brodway House, Carter Lane. London, E. C. 4.

Piaget, J. (1968). *Six psychological studies.* (A. Tenzer, Trans.). New York: Vintage Books. (Original work published 1964).

Piaget, J. (1970a). *Genetic Epistemology.* (E. Duckworth, Trans.), New York: Columbia University Press.

Piaget, J. (1970b). *Science of education and the psychology of the child.* (E. Denoel, Trans.). Grossman Publishers, Inc.

Piaget, J. (1972). *The Principles of Genetic Epistemology.* (W. Mays, Trans.). London: Routledge & Kegan Paul, Ltd.

Piaget, J., & Inhelder, B. (1969a). *The psychology of the child.* (H. Weaver, Trans.). New York: Basic Books. (Original work published 1966).

Piaget, J., & Inhelder, B. (1969b). *The early growth of logic in the child.* (E. Lunzer & D. Papert, Trans.). New York: W. W. Norton & Company, Inc.

Polya, G. (1945) How to solve it. Princeton; Princeton University Press.

Schmuck, R., & Schmuck, P. (1988). *Group processes in the classroom.* (5th ed.). Dubuque, IA: Brown.

Schoenfeld, A. (1985). *Mathematical Problem Solving.* Orlando, FL: Academic Press.

Schoenfeld, A. (Ed.) (1990). *A source book for collegiate mathematics teaching.* MAA Reports, #2.

Schwartz, R. H. (1992). Revitalizing liberal arts mathematics. *Mathematics and computer education,* 26(3), 272–277. (ERIC Document Reproduction Service No. EJ458225)

> Suggests instructional modifications that utilize liberal arts students interests and strengths to stimulate their interest in mathematics. Suggestions recommend that teachers (1) relate mathematics to current critical issues; (2) make use of projects and open-ended problems; (3) utilize collaborative learning; and (4) provide opportunities for written and oral communication.

Schwingendorf, K., Wimbish, G. J., & Hawks-Hoover, J. (1992, January). *An exploratory factor analysis of students enrolled in calculus for management, the social and life sciences.* Paper presented at the Joint Mathematics Meetings of the American Mathematical Society and the Mathematical Association of America, Baltimore.

Schwingendorf, K. & Wimbish, G. J. (1994, January). *Attitudinal changes of calculus students using computer enhanced cooperative learning.* Paper presented at the Joint Mathematics Meetings of the American Mathematical Society and the Mathematical Association of America, Cincinnati, OH.

Serra, M. (1989). Michael Serra's Suggestions on Cooperative Learning. *Cooperative Learning.* Key Curriculum Press, 15–32.

Sharan, S. (1980). Cooperative learning in small groups: Recent methods and effects on achievement, attitudes and ethnic relations. *Review of educational research,* 50, 241–271.

Sharan, S. (Ed.). (1990a). *Cooperative learning: Theory and research.* New York: Praeger.

Sharan, S. (1990b). Cooperative learning and helping behavior in the multiethnic classroom. In H. Foot, M. Morgan, & R. Shute (Eds.), *Children helping children,* 151–176. London: Wiley.

Sharan, S., & Hertz-Lazarowitz, R. (1980). A group-investigation method of cooperative learning in the classroom. In S. Sharan, P. Hare, C. D. Webb, and U. Hertz-Lazarowirz (Eds.), *Cooperation in education.* Provo, UT: Brigham Young University.

Sharan, S., & Hertz-Lazarowitz, R. (1982). Effects of an instructional change program on teachers' behavior, attitudes and perceptions. *Journal of Applied Behavioral Science, 18,* 185–201.

Sharan, S., & Shachar, C. (1988). *Language and learning in the cooperative classroom.* New York: Springer.

Sharan, Y., & Sharan, S. (December 1989/January 1990). Group investigation expands cooperative learning. *Educational Leadership, 47*(4), 17–21.

Sharan, Y., & Sharan, S. (1990). Group investigation expands cooperative learning. *Educational Leadership, 45,* 20–25.

Sharan, Y., & Sharan, S. (1992). *Expanding cooperative learning through group investigation.* New York: Teachers College Press.

Sharan, S., & Shaulov, A. (1990). Cooperative learning, motivation to learn and academic achievement. In S. Sharan (Ed.), *Cooperative learning: Theory and research,* 173–202. New York: Praeger.

Sharan, S., et al, (Eds.) (1980). *Cooperation in education.* Provo, UT: Brigam Young University Press.

Sherman, L. W., & Thomas, M. (January/February 1986). Mathematics achievement in cooperative versus individualistic goal-structured high school classrooms. *Journal of Educational Research, 79,* 169–172.

Skemp, R. R. (1987). *The psychology of learning mathematics* (Revised American Edition). Hillsdale, NJ: Lawrence Erlbaum.

Skinner, C. H. & Smith, E. S. (1992). Issues surrounding the use of self-management interventions for increasing academic performance. *School Psychology Review, 21*(2), 202–210.

Slavin, R. E. (1983a). *Cooperative learning.* New York: Longman.

Slavin, R. E. (1983b). When does cooperative learning increase student achievement? *Psychological Bulletin, 94,* 429–445.

Slavin, R. E. (March 1985a). Cooperative learning: Applying contact theory in desegregated schools. *Journal of Social Issues, 41,* 45–62.

Slavin, R. E. (1985b). Team assisted individualization: A cooperative learning solution for adaptive instruction in mathematics. In M. Qang and H. Walberg, *Adapting instruction to individual differences*, Berkeley, CA: McCutchan.

Slavin, R. E., Sharon, S., Kagen, S., Hertz-Lazarowitz, R., Web, C. & Schmuck, R. (1985). *Learning to Cooperate, Cooperating to Learn.* New York and London: Plenum Publishing.

Slavin, R. E. (1986). *Using student team learning.* Baltimore, Md.: Center for Research on Elementary and Middle Schools, Johns Hopkins University.

Slavin, R. E. (October 1987a). Development and motivational perspectives on cooperative learning: A reconciliation. *Child Development, 58*, 1161–1167.

Slavin, R. E. (1987b). *Cooperative learning: Student teams. What research says to the teacher.* National Education Association, Washington, D. C. (ERIC Document Reproduction Service No. ED282862)

> A review of the research regarding the effectiveness of cooperative learning methods (particularly student teams) indicated that when the classroom is structured in a way that allows students to work cooperatively on learning tasks, students benefit academically as well as socially. The greatest strength of cooperative learning methods is the wide range of positive outcomes that have been found in the research. Cooperative learning methods are usually inexpensive and easy to implement. Teachers need minimal training to use these techniques. The widespread and growing use of cooperative learning techniques demonstrate that, in addition to their effectiveness, they are practical and attractive to teachers. Among such methods are the Student Team Learning procedure (developed at Johns Hopkins University, Maryland); Student Teams Achievement Divisions; Teams Games Tournaments; Team Assisted Individualization; Cooperative Integrated Reading and Composition; Jigsaw; Learning Together; and Group Investigation.

Slavin, R. E. (1988). *Student Team Learning: An Overview and Practical Guide.* National Education Association, Washington, D. C. (ERIC Document Reproduction Service No. ED295910)

> This manual provides descriptions of five cooperative learning methods: (1) Student Teams Achievement Divisions (STAD); (2) Teams Games Tournaments (TGT); (3) Jigsaw; (4) Team Accelerated Instruction (TAI); and (5) Cooperative Integrated Reading and Composition (CIRC). For each of these methods, an overview offers a

description of the procedures followed, how to prepare for it, how to start it, and a schedule of activities involved. Similarities and differences between the methods are discussed and research evidence on the effectiveness of various kinds of team learning is considered. The appendices contain information on scoring methods for different sizes of teams, instructions for making worksheets for team activities, and samples of Jigsaw unit and record forms.

Slavin, R. E. (1990a). Ability grouping, cooperative learning and the gifted. Point-counterpoint-cooperative learning. *Journal for the Education of the Gifted, 14*, 3–8. (ERIC Document Reproduction Service No. EJ420043)

The article discusses how cooperative learning (emphasizing group goals and individual accountability), the limited use of acceleration by extremely able learners, and differentiation within classes can reduce tracking and separate enrichment programs while meeting the needs of gifted students in the regular classroom.

Slavin, R. E. (1990b). Learning together. *American School Board Journal, 177*, 22–23. (ERIC Document Reproduction Service No. EJ413124)

In cooperative learning, students are typically assigned to heterogeneous teams. The Johns Hopkins University Elementary School Program uses four principal methods involving student team learning, including Student Teams Achievement Divisions, Teams Games Tournament, Team Accelerated Instruction Mathematics, and Cooperative Integrated Reading and Composition.

Slavin, R. E. (1990c). Research on cooperative learning: Consensus and controversy. *Educational Leadership, 47*, 52–54. (ERIC Document Reproduction Service No. EJ400501)

Four literature reviews found that cooperative learning methods using group rewards and individual accountability consistently increase student achievement more than control methods in elementary and secondary classrooms. More research is needed to gauge cooperative learning's effectiveness at senior high and college levels and for instilling higher order concepts.

Slavin, R. E. (1990d). *Cooperative learning: Theory, research, and practice.* Englewood Cliffs, N.J.: Prentice-Hall.

Slavin, R. E. (1991a). Are cooperative learning and "untracking" harmful to the gifted? Response to Allan. *Educational Leadership, 48*, 68–71. (ERIC Document Reproduction Service No. EJ422861)

The questions of "untracking" and cooperative learning for the academically gifted are important, because arguments concerning this small population are often used to defeat tracking reduction or elimination efforts for other students. This article rejects ability grouping for high achievers and recommends cooperative programs involving all students.

Slavin, R. E. (1991b). Synthesis of research of cooperative learning. *Educational Leadership, 48*, 71–82. (ERIC Document Reproduction Service No. EJ421354)

For enhancing student achievement, the most successful cooperative learning approaches have incorporated two key elements: group goals and individual accountability. Positive effects have been consistently found on outcomes such as self-esteem, intergroup relations, acceptance of academically handicapped students, attitudes toward school, and ability to work cooperatively.

Slavin, R. E., & Karweit, N. L. (1985). Effects of whole-class, ability grouped, and individualized instruction on mathematics achievement. *American Educational Research Journal, 22*, 351–367.

Slavin, R. E., Leavey,M. B., & Madden, N. A. (1986). *Team accelerated instruction-mathematics.* Watertown, MA.: Mastery Education Corporation.

Smith, E., & Confrey, J. (1991, April). *Understanding collaborative learning: Small group work on contextual problems using a multi-representational software tool.* Paper presented at the Annual Meeting of the American Educational Research Association, Chicago, IL. (ERIC Document Reproduction Service No. ED336391)

The interactions of three high school juniors (two female and one male) working together on a series of contextual mathematics problems using a multi representational software tool were studied. Focus was on determining how a constructivist model of learning, based on an individual problematic-action-reflection model, can be extended to offer explanatory power for small-group collaborative learning. This extension is constructed by adopting several concepts from the socio-historic or Vygotskian school, including the zone of proximal development, cultural tools, proleptic talk, and appropriation. The subjects worked together during a 10-week secondary mathematics course that focused on problem solving with Function Prove. Although constructivist and socio-historic approaches to cognition have, at times, been interpreted as offering opposing viewpoints, it is suggested that there is a potential complementarity, particularly in the area of collaborative peer learning, since researchers in neither area have as yet offered a strong explanatory model for how students jointly construct mathematical knowledge.

Stanford, G. (1977). *Developing Effective Classroom Groups: A Practical Guide for Teachers.* New York: Hart Publishing Company.

Sutton, G. O. (1992). Cooperative learning works in mathematics. *Mathematics Teacher, 85(1)*, 63–66. (ERIC Document Reproduction Service No. EJ 440182)

Shared are personal experiences in the development and implementation of cooperative learning methods in the mathematics classroom. Presented are methods to include greater student participation, interstudent communication, higher-order thinking, and teacher enthusiasm.

Swing, S. R., & Peterson, P. L. (1981, April). *The relationship of student ability and small-group interaction to student achievement.* Paper presented at the Annual Meeting of the American Educational Research Association, Los Angeles, CA. (ERIC Document Reproduction Service No. ED210319)

This study examined student aptitudes and student behaviors during small group interactions as mediators of the effectiveness of small group learning. The hypotheses to be investigated were that the effects of small group learning on student achievement are produced by students' participation in group interaction. All students received regular mathematics classroom instruction. They worked on assignments in mixed ability groups of four students. Achievement, retention and attitude toward mathematics were assessed. A Mann-Whitney comparison showed that trained students participated in more task related interaction than control students. The effects of small group interaction depend on the ability level of the students. Interaction during small group work was most beneficial for low ability students. The study showed they can help themselves by teaching others. A high quality of interaction must prevail if the small group method is to be of maximal effectiveness.

Swing, S. R., & Peterson, P. L. (1982). The relationship of student ability and small-group interaction to student achievement. *American Educational Research Journal, 19*, 259–274.

Tietze, M. (1992). A core curriculum in geometry. *Mathematics Teacher, 85(4)*, 300–303. (ERIC Document Reproduction Service No. EJ446403)

Presents four hands-on activities to teach geometric concepts to both honors and non–college-bound students. The activities are adaptable to various levels and use manipulatives and technology to explore the concepts of enlargement, area, maximum volume, and figure duplication in a cooperative setting.

Towson, S. M. J. (March 1982). Cooperative intervention strategies, perceived status and student self-esteem. *Paper presented at the American Educational Research Association meetings*, New York.

Urion, D. K. & Davidson, N. A. (1992) Student achievement in small-group instruction versus teacher-centered instruction in mathematics. *Primus, 2(3)*, 257–64. (ERIC Document Reproduction Service No. EJ454996)

Reports the results of five contrasts between small-group learning and a more teacher-centered instructional style employed in junior high, high school, and college mathematics classes. In no case did the small-group class perform more poorly than the one led by a teacher; in one case, measuring long-term retention, the small-group class performed better.

Vidakovic, D. (1992, November). *Collaborative work — opportunities for learning through social interaction.* Paper presented at The Fifth Annual International Conference of Technology in Collegiate Mathematics. Rosemont, IL.

Vidakovic, D. (1993). *Cooperative learning: Differences between group and individual processes of construction of the concept of inverse function.* Unpublished doctoral dissertation, Purdue University, Indiana.

Webb, N. M. (1980a). A process-outcome analysis of learning in group and individual settings. *Educational Psychologist, 15*, 69–83.

Webb, N. M. (1980b). An analysis of group interaction and mathematical errors in heterogeneous ability groups. *British Journal of Educational Psychology, 50*, 266–276.

Webb, N. M. (1982a). Student interaction and learning in small groups. *Review of Educational Research, 52*, 421–445. (ERIC Document Reproduction Service No. EJ273691)

Research bearing in three aspects of small group learning is examined: (1) the relationship between interaction and achievement, (2) cognitive process and social-emotional mechanisms bringing interaction and achievement, and (3) characteristics of the individual, group, and reward structure that predict interaction in small groups.

Webb, N. M. (1982b). Group composition, group interaction, and achievement in cooperative small groups. *Journal of Educational Psychology, 74*, 475–484. (ERIC Document Reproduction Service No. EJ267791)

> The relationship between interaction and achievement in cooperative small groups was studied in four junior high school mathematics classrooms. The interaction variable that related most strongly to achievement was asking a question and receiving no response; this type of interaction was negatively related to achievement.

Webb, N. M. (1982c). Peer interaction and learning in cooperative small groups. *Journal of Educational Psychology, 74*, 624–655. (ERIC Document Reproduction Service No.EJ273658)

> The relationship among students and group characteristics, group interaction, and achievement in cooperative small groups were investigated. Three categories of interaction were related to achievement: receiving no explanation in response to a question or error was negatively related to achievement; giving explanations and receiving explanations were positively related to achievement.

Webb, N. M. (1984a). Stability of small group interaction and achievement over time. *Journal of Educational Psychology, 76*, 211–224. (ERIC Document Reproduction Service No. EJ304975)

> This study investigated the stability over time of: (1) student behavior in small groups; and (2) the relationships among student and group characteristics, group interaction, and achievement. Measurements were taken for two three-week instructional units, three months apart, on 110 students in three average-ability junior high school mathematics classrooms.

Webb, N. M. (1984b). Sex differences in interaction and achievement in cooperative small groups. *Journal of Educational Psychology, 76*, 33–44. (ERIC Document Reproduction Service No. EJ304949)

> Using 77 junior high school students in two mathematics classes, this study investigated sex differences in interaction patterns and achievement in small cooperative learning groups. Results related to the male-female ratio in a group. Explanations for these results and consequences for group composition in the classroom are discussed.

Webb, N. M. (1984c). Microcomputer learning in small groups: Cognitive requirements and group processes. *Journal of Educational Psychology, 76*, 1076–1088. (ERIC Document Reproduction Service No. EJ310883)

This study investigated the cognitive abilities, cognitive styles, and student demographic characteristics that predicted learning of computer programming in small groups; the group process variables that predicated learning of computer programming; and the student characteristics that related to group processes. Different profiles of abilities predicted different programming outcomes.

Webb, N. M. (1985a). The role of gender in computer programming learning processes. *Journal of Educational Computing Research, 1*, 441–458. (ERIC Document Reproduction Service No. EJ327051)

Two studies investigated gender differences in planning and debugging behavior, group processes, and achievement among junior high school students learning either Logo or BASIC computer programming in small groups or as individuals. Results indicate no differences in behavior and achievement of males and females.

Webb, N. M. (1985b). Verbal interaction and learning in peer-directed groups. *Theory into Practice, 24*, 32–39. (ERIC Document Reproduction Service No. EJ317595)

This article describes: (1) the small group settings and peer group research; (2) verbal interaction and its benefits or detriments for learning; (3) students' cognitive abilities and demographic characteristics that predict their participation in small group interaction; (4) group composition and group interaction; and (5) the teacher's role in promoting group interaction.

Webb, N. (1985c). Cognitive requirements of learning computer programming in group and individual settings. *AEDS Journal, 18*, 183–194. (ERIC Document Reproduction Service No. EJ317206)

This study compared achievement of students learning programming in pairs and demographics. No difference were found in programming outcome, but mathematics and verbal ability best predicted individual setting outcomes, while nonverbal reasoning, spatial ability, and age best predicted learning in groups.

Webb, N. M. (1991). Task-related verbal interaction and mathematics learning in small groups. *Journal for Research in Mathematics Education, 22*, 366–389.

Webb, N. M., & Cullian, L. K. (1981, August). *Group process as the mediator between aptitudes and achievement: Stability over time.* Paper presented at the Annual Convention of the American Psychological Association, Los Angeles, CA. (ERIC Document Reproduction Service No. ED209601)

> Most research on small group learning has focused on achievement, but few studies have systematically investigated the effects of group processes on achievement, or the influences of individual aptitudes and group composition on group processes. To investigate the relationship between student aptitudes, group process, and achievement in cooperative small groups in junior high school mathematics classrooms, and the stability of the relationship over time, students ($n = 105$) in four classrooms participated in two studies. Initially, all students learned a one-week unit on consumer mathematics. Three months later, half of the students learned a one-week unit on probability. Students worked in four-person homogeneous-ability or heterogeneous-ability groups; they also completed achievement tests and the Eysenck Personality Inventory. Analyses of the data indicated that group process was a potent predictor of achievement in all studies; "asking a question and receiving no answer," the best predictor of achievement, was detrimental to achievement. The effects of group composition and student aptitudes on achievement were mediated by the group process variable. The mediating effect of group process and the magnitude of coefficients were stable across studies; group process was stable over time, both in average frequency and in individual student levels of participation. The results suggest a need to determine whether the stability of group process is generalizable to longitudinal designs.

Webb, N. M., & Cullian, L. K. (1983). Group interaction and achievement in small groups: Stability over Time. *American Educational Research Journal, 20*, 411–423. (ERIC Document Reproduction Service No. EJ289078)

> The relationships among (a) student and group characteristics, group interaction, and achievement in small groups in junior high school mathematics classrooms and (b) the stability of these relationships over time were investigated. Interaction in the group was a potent predictor of achievement.

Webb, N. M., & Kenderski, C. M. (1984). Student interaction in small group and whole class settings. In P. L. Peterson, L. C. Wilkinson, & M. Hallinan (Eds.), *The social context of instruction: Group organization and group processes.* New York: Academic Press.

Webb, N. M., & Kenderski, C. M. (1985). Gender-related differences in small group interaction and achievement in high-achieving and low-achieving classrooms. In L. C. Wilkinson & C. B. Marrett (Eds.). *Gender-related differences in classroom interaction,* 209–236. New York: Academic Press.

Webb, N. M., Ender. P. & Lewis, S. (1986). Problem solving strategies and group processes in small groups learning computer programming. *American Educational Research Journal, 23,* 243–261. (ERIC Document Reproduction Service No. EJ351723)

> Planning and debugging strategies and group processes predicating learning of computer programming were studied in 30 students aged 11 to 14. Students showed little advance planning. Factors associated with learning included high-level planning of program chunks, debugging of single statements, explaining, and verbalizing aloud while typing.

Weissglass, J. (1985). *Exploring elementary mathematics: A small-group approach for teaching.* Santa Barbara, CA.: Kimberly Press.

Weissglass, J. (1987). *Learning, feelings, and educational change, (Part I),* Santa Barbara, CA.: Kimberly Press.

Weissglass, J. (1990). Cooperative learning using a small-group laboratory approach. In N. Davidson *Cooperative learning in mathematics: A handbook for teachers.* Menlo Park, CA: Addison-Wesley Publishing Co.

Wilkinson, L. C. (1985). Communication in all-student mathematics groups. *Theory Into Practice, 24,* 8–13.

Wilkinson, L. C., Lindow, J., & Chiang, C. P. (1985). Sex differences and sex segregation in students' small-group communication. In L. C. Wilkinson & C. B. Marrett (Eds.), *Gender influences in classroom interaction,* 185–207. New York: Academic Press.

Wimbish, G. J. (1993). Identification and classification of attitudes of non-specialist undergraduate mathematics students that might affect collegiate cooperative learning procedures (Doctoral Dissertation, The University of Alabama). To appear.

Wimbish, G. J. (1990, November). *A comparison of the effects of two oral protocol interventions on college students' analytic skill: A pilot study.* A paper presented at the annual MSERA meeting. New Orleans.

Wood, J. B. (1992). *The application of computer technology and cooperative learning in developmental algebra at the community college.* Paper presented at the Annual Computer Conference of the League for Innovation in the Community College (9th, Orlando, FL, October 21–24). (ERIC Document Reproduction Service No. ED352099)

> In fall 1991, a study was initiated at Central Florida Community College (CFCC) in Ocala to examine the effects of computer lab tutorials and cooperative learning on mathematics achievement, retention rate, mathematics anxiety, mathematical confidence, and success in future mathematics courses among 29 students in an intermediate algebra class. Another course section of 23 intermediate algebra students, taught by the same instructor but utilizing the traditional lecture method, served as a control group. The experimental section was divided into groups of two to four students having similar achievement placement test scores. Homework assignments, computer lab tutorials, and all tests (except for the final exam) were completed on a group basis, with issues of assignment and lab meeting times, group participation guidelines, and class attendance decided and monitored by the group. Both classes were given the Fennema-Sherman Mathematics Anxiety and Confidence Scales test before and after the course. Study findings included the following: (1) a total of 23 students in the experimental group, and 15 students in the control group completed the course; (2) the control group showed greater increases in postcourse confidence ratings and greater reductions in anxiety ratings than the experimental group; (3) 69% of experimental group students received a course grade of A, B, or C, as compared with 52% of the control group; and (4) 87.5% of control group students were successful in their subsequent math course compared to 80% of the experimental group students. Data tables, and narrative excerpts of midterm and final written evaluations by students in the experimental group are included.

Workman, M. B. (1990). *The effects of grouping patterns in a cooperative learning environment on student academic achievement.* Unpublished master's thesis, Dominican College. (ERIC Document Reproduction Service No. ED319617)

> This study investigated the effects of grouping patterns in a cooperative learning environment on the mathematics achievement of two high school geometry classes. Heterogeneous, homogeneous, random, mixed gender and single gender grouping patterns were coordinated with six similar geometry topics and students were grouped accordingly for a period of 10 weeks. The t test was applied to pretests and posttests after each grouping pattern and only two grouping patterns showed significant results. However, the means, standard deviations, teacher observations, and student responses revealed other implica-

tions in relation to grouping and a favorable response to cooperative small group work.

Yackel, E., Cobb, P., & Wood, T. (1991). Small-group interactions as a source of learning opportunities in second-grade mathematics. *Journal for Research in Mathematics Education, 22*(5), 390–408.

Index

accommodation, Piaget's theory, 20
achievement divisions, 85
ACT scores, 25, 27
activities, 31, 32, 38, 45, 52, 53
    examples, 33
    purpose, 31, 32
adult students, 1, 28
algebra
    intermediate, 37, 43, 48
    linear, 35, 37, 41, 44, 47, 48
Aronson, 88
assessment, 55, 63, 64, 92
    as a motivator, 55, 66
    class participation form, 7, 125
    grading schemes, 66, 131
    group test, 57, 58, 60–62
    individual test, 59
    of all aspects of class work, 55,
        64–66
    participation in class tasks, 64,
        75
    quiz, 60
    tests, examinations, 107, 108
assignment to groups, permanent, 4,
    5, 8, 29
assimilation, Piaget's theory, 20, 43
attitudes
    about mathematics, 12, 21
    change in attitude, 13, 78

Balacheff, 18
base groups, 29, 89, 92
basis of vector space, 37, 44, 48

beliefs, about mathematics, 12, 13,
    21
Black Issues in Higher Education, 1

calculus, 34, 40, 46, 60, 61
    grading schemes, 131, 133
class participation
    evaluation of, 5, 7, 64–66, 125,
        131–134
    on syllabus, 127–129
class participation form, 125
class tasks, 31, 38, 45, 50, 53
    examples, 40
Cobb, 18
cognitive development, 20
cognitive differences, 20
collective knowledge of group, 4, 55
communication, need to be under-
    stood, 20
commuters, 24, 26–28
composition of groups, 25, 29
computer
    as additional group member, 17,
        49
    laboratory, 49
    programs, use of, 16, 45
    use of, 35, 37, 38, 47
computer algebra system, 33, 37, 38,
    49, 52, 113
conceptual
    development, 20
    understanding, 38, 50, 78
conferences with students, 109, 111